STRATHCLYDE UNIVERSITY LIBRARY
30125 00218751 5

D0121463

Receptors and Recognition

General Editors: P. Cuatrecasas and M.F. Greaves

About the series

Cellular Recognition – the process by which cells interact with, and respond to, molecular signals in their environment – plays a crucial role in virtually all important biological functions. These encompass fertilization, infectious interactions, embryonic development, the activity of the nervous system, the regulation of growth and metabolism by hormones and the immune response to foreign antigens. Although our knowledge of these systems has grown rapidly in recent years, it is clear that a full understanding of cellular recognition phenomena will require an integrated and multi-disciplinary approach.

This series aims to expedite such an understanding by bringing together accounts by leading researchers of all biochemical, cellular and evolutionary aspects of recognition systems. This series will contain volumes of two types. First, there will be volumes containing about five reviews from different areas of the general subject written at a level suitable for all biologically oriented scientists (Receptors and Recognition, series A). Secondly, there will be more specialized volumes (Receptors and Recognition, series B), each of which will be devoted to just one particularly important area.

Advisory Editorial Board

Receptors and Recognition

Series A

Published

Volume 1 (1976)

M.F. Greaves (London), Cell Surface Receptors: A Biological Perspective
F. Macfarlane Burnet (Melbourne), The Evolution of Receptors and Recognition in the Immune System
K. Resch (Heidelberg), Membrane Associated Events in Lymphocyte Activation
K.N. Brown (London), Specificity in Host-Parasite Interaction

Volume 2 (1976)

D. Givol (Jerusalem), A Structural Basis for Molecular Recognition: The Antibody Case
B.D. Gomperts (London), Calcium and Cell Activation
M.A.B. de Sousa (New York), Cell Traffic
D. Lewis (London), Incompatibility in Flowering Plants
A. Levitski (Jerusalem), Catecholamine Receptors

Volume 3 (1977)

J. Lindstrom (Salk, California), Antibodies to Receptors for Acetylcholine and other Hormones
M. Crandall (Kentucky), Mating-Type Interaction in Micro-organisms
H. Furthmayr (New Haven), Erythrocyte Membrane Proteins
M. Silverman (Toronto), Specificity of Membrane Transport

Volume 4 (1977)

M. Sonenberg and A.S. Schneider (New York), Hormone Action at the Plasma Membrane: Biophysical Approaches
H. Metzger (NIH, Bethesda), The Cellular Receptor for IgE
T.P. Stossel (Boston), Endocytosis
A. Meager (Warwick) and R.C. Hughes (London), Virus Receptors
M.E. Eldefrawi and A.T. Eldefrawi (Baltimore), Acetylcholine Receptors

Volume 5 (1978)

P.A. Lehmann (Mexico), Stereoselective Molecular Recognition in Biology
A.G. Lee (Southampton, U.K.), Fluorescence and NMR Studies of Membranes
L.D. Kohn (NIH, Bethesda), Relationships in the Structure and Function of Receptors for Glycoprotein Hormones, Bacterial Toxins and Interferon

Volume 6 (1978)

J.N. Fain (Providence, Rhode Island), Cyclic Nucleotides
G.D. Eytan (Haifa) and B.J. Kanner (Jerusalem), Reconstitution of Biological Membranes
P.J. O'Brien (NIH, Bethesda), Rhodopsin: A Light-sensitive Membrane Glycoprotein

Index to Series A, Volumes 1–6

Series B

Published

The Specificity and Action of Animal Bacterial and Plant Toxins (B1)
edited by P. Cuatrecasas (Burroughs Wellcome, North Carolina)
Intercellular Junctions and Synapses (B2)
edited by J. Feldman (London), N.B. Gilula (Rockefeller University, New York) and J.D. Pitts (University of Glasgow)
Microbial Interactions (B3)
edited by J.L. Reissig (Long Island University, New York)
Specificity of Embryological Interactions (B4)
edited by D.R. Garrod (University of Southampton)
Taxis and Behavior (B5)
edited by G.L. Hazelbauer (University of Uppsala)
Bacterial Adherence (B6)
edited by E.H. Beachey (Veteran's Administration Hospital and University of Tennessee, Memphis, Tennessee)
Virus Receptors Part 1 Bacterial Viruses (B7)
edited by L.L. Randall and L. Philipson (University of Uppsala)
Virus Receptors Part 2 Animal Viruses (B8)
edited by K. Lonberg-Holm (Du Pont, Delaware) and L. Philipson (University of Uppsala)
Neurotransmitter Receptors Part 1 Amino Acids, Peptides and Benzodiazepines (B9)
edited by S.J. Enna (University of Texas at Houston) and H.I. Yamamura (University of Arizona)
Neurotransmitter Receptors Part 2 Biogenic Amines (B10)
edited by H.I. Yamamura (University of Arizona) and S.J. Enna (University of Texas at Houston)
Membrane Receptors: Methods for Purification and Characterization (B11)
edited by S. Jacobs and P. Cuatrecasas (Burroughs Wellcome, North Carolina)
Purinergic Receptors (B12)
edited by G. Burnstock (University College, London)
Receptor Regulation (B13)
edited by R.J. Lefkowitz (Duke University Medical Center, North Carolina)
Histocompatibility Antigens: Structure and Function (B14)
edited by P. Parham (Stanford University School of Medicine, California) and J. Strominger (Harvard University, Massachussetts)
Receptor-Mediated Endocytosis (B15)
edited by P. Cuatrecasas (Burroughs Wellcome, North Carolina) and T. Roth (University of Maryland, Baltimore County)

Receptors and Recognition

Series B Volume 16

Genetic Analysis of the Cell Surface

Edited by
P. Goodfellow

Imperial Cancer Research Fund Laboratories, London, U.K.

LONDON NEW YORK
CHAPMAN AND HALL

First published 1984 by
Chapman and Hall Ltd
11 New Fetter Lane, London EC4P 4EE

Published in the USA by
Chapman and Hall
733 Third Avenue, New York NY 10017

Printed in Great Britain at the
University Press, Cambridge

ISBN 0 412 25070 5

British Library Cataloguing in Publication Data

Genetic analysis of the cell surface.—
(Receptors and recognition; B16)
1. Cell membranes 2. Mammals—Cytology
3. Genetics
I. Goodfellow, Peter II. Series
599.08′76 QH601

ISBN 0–412–25070–5

Library of Congress Cataloging in Publication Data

Main entry under title:

Genetic analysis of the cell surface.
(Receptors and recognition. Series B; v. 16)
Includes bibliographical references and index.
1. Plasma membranes. 2. Cytogenetics. I. Goodfellow, P. (Peter), 1951– . II. Series.
[DNLM: 1. Cell membrane—Physiology. 2. Cell membrane—Ultrastructure.
3. Genetic technics. W1 RE107MA v. 16/QH 601 G3265]
QH601.G43 1984 574.87′5 84–1758

ISBN 0–412–25070–5

Contents

Contributors

Peter W. Andrews, The Wistar Institute of Anatomy and Biology, 36th Street at Spruce, Philadelphia 19104, U.S.A.

Walter F. Bodmer, Imperial Cancer Research Fund, Lincoln's Inn Fields, London WC2A 3PX, U.K.

Robert P. Erickson, Department of Human Genetics, University of Michigan School of Medicine, 1137 E. Catherine Street, Ann Arbor, Michigan 48109, U.S.A.

Jeffrey A. Frelinger, Department of Microbiology and Immunology, University of North Carolina Medical School, Chapel Hill, North Carolina 27514, U.S.A.

Peter Goodfellow, Imperial Cancer Research Fund, Lincoln's Inn Fields, London WC2A 3PX, U.K.

Janet Lee, Imperial Cancer Research Fund, Lincoln's Inn Fields, London WC2A 3PX, U.K.

Melanie J. Palmer, Department of Microbiology and Immunology, University of North Carolina Medical School, Chapel Hill, North Carolina 27514, U.S.A.

Nobuyoshi Shimizu, Department of Molecular and Cellular Biology, University of Arizona, Tucson, Arizona 85721, U.S.A., and Department of Molecular Biology, Keio University School of Medicine, 35 Shinanomachi, Shinjuku-ku, Tokyo, Japan.

Patricia Tippett, MRC Blood Group Unit, University College London, U.K.

John Trowsdale, Imperial Cancer Research Fund, Lincoln's Inn Fields, London WC2A 3PX, U.K.

Alan Tunnacliffe, Imperial Cancer Research Fund, Lincoln's Inn Fields, London WC2A 3PX, U.K.

Preface

The cell surface is the barrier between the cell and its environment which regulates the flow of both simple and complex molecules into and out of the cell; it is also the organelle responsible for communication between the cell and its environment. Each cell expresses receptors for a wide variety of hormones, growth factors, growth substrates and other cells. In multicellular organisms communication between cells is required for controlling development, cellular differentiation, morphogenesis and, in a more general sense, integration of myriad cell types into a single organism. The series *Receptors and Recognition* has as its overall aim the dissection of the cell surface to correlate structure and function for this complex organelle. In most of the preceding volumes the approach has been biochemical or physiological. In this volume the mammalian cell surface is analysed by a genetic approach.

Genetic analysis of the cell surface, especially when combined with immunological techniques, has a long history. In 1900 Landsteiner showed that serum from one individual could agglutinate the red cells of another. Besides the practical result of making blood transfusion safe, this was the first demonstration of a human genetic polymorphism and for the next 50 years the red blood cell surface provided most of the genetic markers used to study human populations. The genetics of the surface of nucleated cells can be traced back to early experiments in cancer research and attempts to learn the rules associated with tumour transplantation. The transfer of tumour cells from one mouse to another could be used to define mice which were apparently resistant to tumour killing; this resistance was clearly very complicated and depended both on the tumour transferred and the recipient mouse. The problem was eventually solved by using two new genetic creations: the inbred mouse and the congenic mouse (see Chapter 1). Snell, in one of the most determined series of experiments in modern times (the experiments took several years and had to be started twice because of a catastrophic fire), produced mice which differed at a single region which controlled tumour transplantation: congenic mice which differed at this region rejected tumours exchanged between them; mice which were identical at this region accepted tumours. This was the first genetic demonstration of the mouse major histocompatibility region or complex (MHR or MHC) to which Snell gave the prosaic name *H-2* (histocompatibility locus-2). Following the prescient suggestions of Landsteiner that transplantation and blood transfusion reactions would have a similar basis, Gorer made antisera which could predict the outcome of tumour transplantation experiments. These antibodies recognized products of the *H-2* locus. Thus, the surface of tumour cells contains genetically regulated molecules which are recognized as foreign by recipient mice. Medawar, studying skin transplantation in humans, demonstrated that normal tissues also expressed genetically controlled histocompatibility antigens and this was the stimulus which eventually led to the definition of the human MHC or *HLA* complex.

Subsequent advances in immunological and genetic techniques have been applied singularly and together for analysis of the cell surface. Particularly important were the introduction of somatic cell genetics (see Chapter 3), monoclonal antibodies (see Chapters 1 and 3) and the new techniques of DNA manipulation (see Chapter 4). Today genetic analysis of the cell surface is applied at many different levels from the structure of populations to the primary DNA sequence. In population studies departures from expected gene frequencies for cell surface markers can give indications of selection which in turn may give intimations of function. Similarly, at the level of the whole organism individuals lacking particular antigens may show increased or decreased abilities to cope with physiological stress or infection. At the cellular level interactions can be investigated by studying communication between cells with different cell surface phenotypes or by blocking interactions with specific antibodies. Also at the cellular level mutants can be selected *in vitro* which lack functional receptors and this can be correlated with changes in cellular behaviour. Gene mapping or genetic analysis at the chromosomal level is a *sine qua non* for all genetic analysis and can provide information about functional and evolutionary relationships between genes in the same and different systems. Finally, in the DNA sequence is the answer to many of the questions we would like to ask, if we but knew how to interpret the information.

The chapters in this volume are designed to illustrate the wide variety of methods available for the genetic analysis of the cell surface. For reasons of space, consideration has been largely limited to mice and men but similar studies have also proved fruitful in other mammals, simpler eukaryotes and prokaryotes.

1 Immunogenetic Approaches to Cell Surface Molecules in the Mouse

MELANIE J. PALMER and JEFFREY A. FRELINGER

EDITOR'S INTRODUCTION

The genetics of both mice and men have been extensively studied and in both cases cell surface molecules have proved to be amenable to genetic analysis. However, studying mice has one great advantage: mice are experimental animals and mice with specific desired genetic constitutions can be created to solve particular problems. Besides being able to mate mice with required characteristics it is also possible to reduce genetic complexity by using inbred mice and congenic strains of inbred mice.

The combination of immunological and genetic techniques has been termed immunogenetics. As first demonstrated by Gorer, Lyman and Snell for *H-2*, employing antibodies to define cell surface molecules and congenic mice to define the genetic loci involved is a very powerful approach. In recent years the immunogenetic approach has been used to fine map *H-2* and has been extended to many other systems. Particularly fruitful has been the analysis, instituted by Boyse, of the cells which interact to form the immune system. The definition of T lymphocytes, B lymphocytes, subsets of T and B lymphocytes and even lymphocyte precursors and different maturation stages have all been facilitated by immunogenetics. In Chapter 1 Melanie Palmer and Jeffrey Frelinger explain the basic techniques of murine immunogenetics and stress the use of the two relatively new methods of recombinant inbred strains of mice and monoclonal antibodies. The principles involved are illustrated by reference to two antigenic systems: β_2-microglobulin, which is found on all cells, and Ly-5, a differentiation antigen which is restricted to lymphocytes.

Acknowledgements

We thank Jean Holliday and Joyce Bradshaw for excellent and patient secretarial assistance. J.A.F. is the recipient of an American Cancer Society Faculty Research Award.

Genetic Analysis of the Cell Surface
(*Receptors and Recognition*, Series B, Volume 16)
Edited by P. Goodfellow
Published in 1984 by Chapman and Hall, 11 New Fetter Lane, London EC4P 4EE

1.1 INTRODUCTION

Our purpose in writing this chapter is to demonstrate the use of two distinct, immunological approaches to the study of cell surface molecules. These approaches are (1) the production and use of genetically standardized inbred mouse strains and (2) the use of antibodies produced either by immunization between genetically defined individuals or by myeloma–B cell fusions (monoclonal antibodies). The inbred mouse strains provide a constant pool of genes coding for cell surface molecules and the antibodies provide tools for direct access to the molecules themselves. These reagents can be employed together to provide structural as well as genetic information about cell surface molecules.

Congenic mouse strains (those differing at only a small number of genes on a given chromosome) simplify antisera production by genetically limiting the possible antigenic differences to those coded by a small section of a single chromosome. Alloantisera are produced by immunization of one strain with tissues of another strain. The use of congenic mouse strains in conjunction with analysis of alloantisera can define new antibodies which can then be used to identify novel cell surface components.

It is interesting that at the time of inception of congenic mouse strains, their utility was not generally appreciated. When George Snell began the production of H-2 congenics he was thought to have developed a tool of little use to any investigators outside of a rather esoteric branch of transplantation biology (Snell, 1958). In the 1970s it became increasingly apparent that such animals were of great utility for the understanding of the interaction of molecules on the cell surface that were involved in regulation of immune responses (Benacerraf and McDevitt, 1972). Although initially used to study major histocompatibility genes, this approach is now used to study other genes such as immunoglobulin and minor histocompatibility genes. The immunogenetic approach has gone from being one which was rather arcane to one which is now a straightforward, accepted approach to solving almost any problem in cell surface biochemistry. This approach has been especially useful for investigating the immune system.

In addition to congenic mice, another genetic approach, again not appreciated at the time of its inception, has been useful. This is the production of recombinant inbred strains of mice (RI strains) (Bailey, 1971). Although RI strains have not yet affected immunobiology as much as the congenics, they have begun to have an impact on the genetics of cell surface molecules.

It has been over thirty years since Gorer, Lyman and Snell were able to show that the gene for blood group 2 in the mouse was the gene responsible for tumour rejection between inbred strains of mice (Gorer *et al.*, 1948). This identity provided the first evidence that there were links between genes involved in cell surface phenomena such as graft rejection, and those detected by antibodies. Over the years, serological approaches have been used for the study of many cell surface-mediated events. In addition, methods such as affinity chromatography and antibody inhibition studies can provide information about the biochemistry and function of cell surface molecules. Antibodies can also be used to interfere with physiological functions. For example, in experimental myasthenia gravis antibodies are produced which block the acetylcholine receptor, resulting in blockage of nerve–muscle signal transmission. In turn, antibodies can be produced which inhibit the antibodies blocking the acetylcholine receptor. These antibodies, since they can block the binding of the first antibody to the acetylcholine receptor, might be used as a therapeutic agent to treat patients with myasthenia gravis. Antibodies have also been used to determine the function of cells involved in the immune response. For instance, antibodies directed at Ia molecules have been used to block immune responses by interfering with antigen presentation on the surface of macrophages. This inhibits the recognition of antigen–Ia complexes by T cells.

The utility of both antibodies and congenic mice is not in their separate use but rather their use in conjunction to define cell surface molecules. Complex antisera can be analysed using congenic mouse strains which express different allelic forms of cell surface molecules. Through the absorption of antibody molecules from the complex antisera with a series of different congenic mouse cells, an antiserum that is specific for one antigen can be produced. Furthermore, by cross-immunizing with congenic mouse strains the complexity of the antibody specificities within the sera can be greatly reduced. By carefully choosing strains which share many known antigens, antibodies can be produced to the small number of antigenic differences between the donor and recipient. The extraordinary usefulness of congenic mice in the production of alloantisera lies in the ability to produce alloantisera specific for a set of allelic cell surface molecules.

We have chosen to discuss two murine examples in which congenic and serological analysis have been used to study cell surface molecules. These are β_2-microglobulin and Ly-5/T200.

The first example, β_2-microglobulin (β_2m), was first described in human urine as a free protein and only later described as a human and mouse cell surface component (Berggard and Bearn, 1968). Initially, the structural resemblance of β_2m to one of the constant domains of immunoglobulins led investigators to speculate that it represented a free immunoglobulin domain (Smithies and Poulik, 1972; Peterson *et al.*, 1972). When it was discovered that β_2m occupied an important role in the expression of major histocompatibility complex (MHC) determinants (Poulik *et al.*, 1974), there was much excitement over β_2m as an evolutionary link between the cellular and humoral arms of the immune system. However, genetics of β_2m and the cell surface functions have remained elusive. Only recently in mice has its genetic location been discovered through the use of congenic and RI strains in combination with antisera produced previously.

In 1975, two groups, searching for T cell differentiation antigens, inadvertently discovered the same molecule by different routes, and named it Ly-5 and T200 (Trowbridge *et al.*, 1975; Komuro *et al.*, 1975). It was the development of a T200-specific monoclonal that finally allowed investigators to demonstrate that Ly-5 and T200 represent the same molecule (Trowbridge, 1978). These molecules are members of a multigene family and members of the same gene family are expressed on most haematopoietic cells. The Ly-5/T200 studies demonstrate the utility of combining the use of congenic mouse strains with serological techniques in characterizing cell surface molecules.

1.2 BACKGROUND TO IMMUNOGENETICS AND THE MAJOR HISTOCOMPATIBILITY COMPLEX

One way of gaining further insight into immunological recognition events is to characterize molecules on the surface of lymphoid cells by biological and chemical means. Another method of dissecting these events at the molecular level is to study the genetics of the molecules involved in mediating immunological phenomena. Immunogenetics involves locating and studying the genes which code for molecules involved in immunological phenomena. One cluster of genes exhaustively studied by immunogenetics is the major histocompatibility complex (see Chapter 4). These gene clusters are designated *HLA* in man and *H-2* in the mouse. The *H-2* cluster consists of a series of linked genes located on mouse chromosome 17 (Fig. 1.1). It is composed of three major regions which are further divided into subregions. The *K* and *D* regions control cytotoxic cellular alloantigens involved in tissue transplantation. They are referred to as class I molecules. Class II genes code for products which are chemically and functionally distinct from class I gene products. The class II

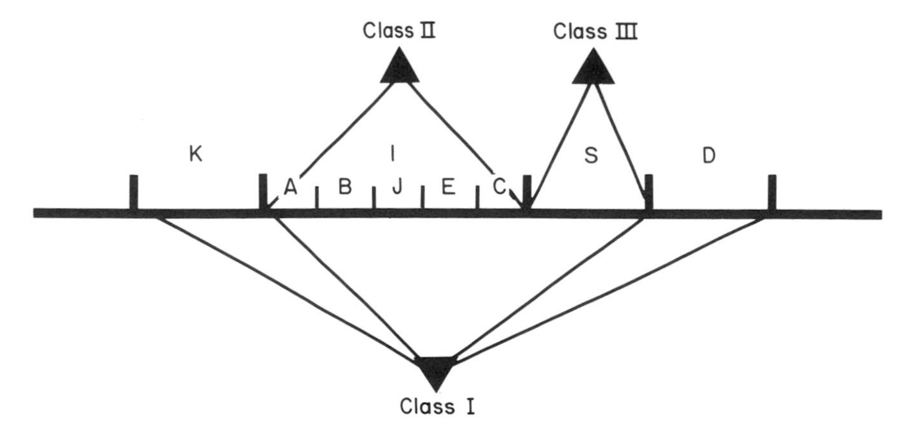

Fig. 1.1 The major histocompatibility complex of mouse.

gene products map to the *I* region. They are designated Ia antigens, and are involved in the control of immune responses. Class III genes located in the *S* region contain the structural gene for C4, a complement component. Each of the regions of the MHC is multiallelic. Polymorphism, or the genetic differences between alleles, provides the basis for such phenomena as graft rejection and mixed lymphocyte responses.

1.3 DEFINITION OF CONGENICS

Congenic mice provide a powerful tool for studying the location and function of both minor and major histocompatibility genes. Two mice which are identical at all but a single locus are said to be co-isogenic. Co-isogenicity is extremely hard to achieve experimentally. It occurs only by a point mutation arising in an inbred strain. However, mice which differ at a small portion of chromosome can be produced by extensive backcrossing. Such pairs of strains are termed congenic. An example of a congenic strain pair would be two strains which differ only at their MHC loci and share otherwise identical (background) genes. The combination of all the alleles at loci of interest on a single chromosome is referred to as the haplotype. The haplotype is denoted with a letter superscript. For instance, strains B10 and B10.A have identical background genes of the B10 strain. Strain B10.A has the haplotype $H\text{-}2^a$ of the A strain MHC, while B10 has the haplotype $H\text{-}2^b$. Congenic mice are important when studying a given gene because the elimination of unlinked genetic factors can substantially decrease experimental error. For example, when cells of B10.A are mixed in tissue culture with irradiated cells of B10 (called stimulators), they will be stimulated to divide and mature into killer cells by foreign antigens (alloantigens) expressed on the surface of B10 cells. Because both strains are identical except at the MHC, the genes coding for the foreign antigens recognized by B10.A must map within the MHC. By using strains which share some MHC determinants by descent, it is possible to assign functions to smaller

Table 1.1 B10 congenic mouse strains*

	K	Aα	Aβ	Eβ	J	Eα	S	D
B10	b	b	b	b	b	b	b	b
B10.A	k	k	k	k	k	k	d	d
B10.K	k	k	k	k	k	k	k	k
B10.D2	d	d	d	d	d	d	d	d
B10.P	p	p	p	p	p	p	p	p

* Each of the above congenic strains was derived from the background strain C57BL/10 (B10).

genetic regions. Strain B10.A shares the *K* end of *H-2* with B.10K and the *D* end with B10.D2. Using B10.A cells in combination with B10.K or B10.D2 cells allows the assignment of genes expressed on the stimulator cells to be mapped to specific regions within the MHC (Table 1.1).

1.3.1 Production of congenic lines

Mice which are bred randomly without regard to their genetic relationships will be heterozygous at many loci. Inbreeding or mating of mice genetically related to one another (usually brother–sister matings) restricts genetic heterogeneity because as inbreeding continues segregating alleles become fewer. The series of genetic crosses leading to the production of a congenic mouse strain always begins with two strains. One strain, referred to as the background strain, provides the genetic background and must be inbred to provide a uniform genetic constitution. The other strain (donor strain) donates the differential locus or loci and need not be inbred.

While there are several methods for producing congenic mice differing at the

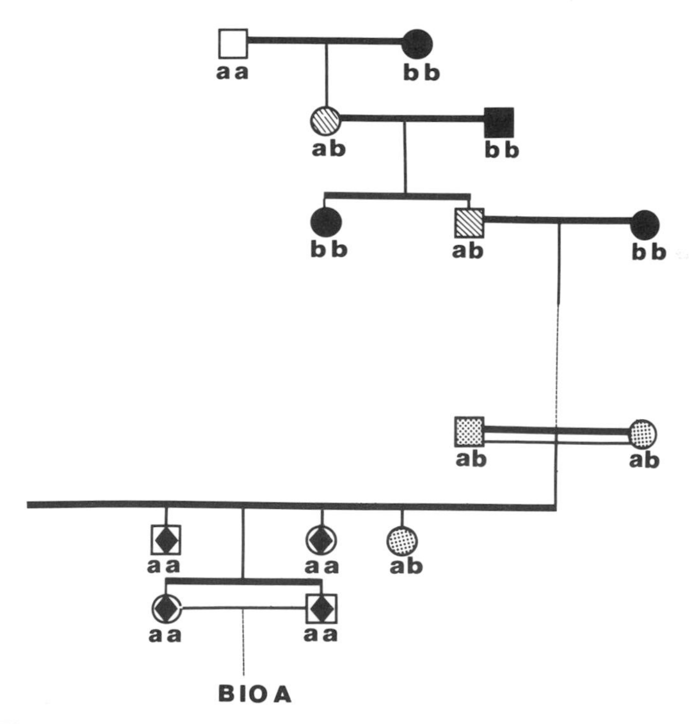

Fig. 1.2 The NX backcross system – explanation in text; aa = strain A; bb = strain B10 (C57BL/10).

H-2 locus (Snell, 1948; Green and Doolittle, 1963; Green, 1966; Snell and Bunker, 1965), the simplest system is the NX backcross system (Fig. 1.2). This system requires that the selected trait is detectable in heterozygotes. Hence this system cannot be used to produce congenics of recessive traits. Fortunately, most genes encoding cell surface molecules are expressed co-dominantly, and are easily detectable in heterozygotes. The production of the congenic line begins with crossing the two strains and 'backcrossing' the hybrid mouse to the parental background strain. The N_1 generation (identical with F_1) is heterozygous, having inherited half of its genetic material from parent A and half from parent B. The N_1 generation is backcrossed to the background parent. The resulting segregating N_2 generation is tested for the desired trait. N_2 animals possessing the desired donor trait (in this case H-2) are again backcrossed to the background parental strain and the N_3 progeny are tested for retention of the donor trait. This process is repeated for a minimum of 12 backcross generations at which time the heterozygous mice are intercrossed and mice which are homozygous for the donor trait are selected and maintained by brother–sister matings as inbred strains. The 12 backcross generations reduce the possibility of an unlinked (and undesired) gene being fixed in the congenic strain to less than 0.01%. To borrow an analogy from J. Klein, producing a congenic strain is like producing the perfect martini. If you start with a martini one part gin to one part vermouth (with an olive) it is not dry enough. Pour out half (save the olive!) and add more gin. If it is still not dry enough repeat 10 more times and the perfect martini results. It is 99.97% gin and 0.03% vermouth and still contains the olive. The olive represents the selected gene, the vermouth represents the donor strains genetic background and the gin represents the congenic partner. The rationale is to replace the donor strain genes of the N_1 heterozygote with background genes while selecting to retain one specific genetic trait derived from the donor strain. An example of congenic lines is illustrated in Table 1.1. All strains listed have identical B10 backgrounds but have been constructed to have different *H-2* genes.

Another genetic approach to studying the location of the genes coding for cell surface components is the use of recombinant inbred (RI) strains (Bailey, 1971). RI strains are a group of inbred strains derived from a cross between two different inbred strains. The F_1 mice from the first cross are paired at random and in subsequent brother–sister matings there is a unique pattern of fixation of alleles in each strain. Each genetic locus will have a particular pattern of fixation among the strains of the RI panel. Each locus in which the two parental strains are distinguishable will have a characteristic distribution pattern of alleles in the RI strains. This pattern is called a Strain Distribution Pattern (SDP). RI strains are useful in large panels where the distribution of a particular allele among the panel of strains provides a way to study segregation and interaction among genes.

1.4 SEROLOGICAL ANALYSIS

A heterogeneous population that consists of genetically unrelated individuals will share some antigens and not others. Immunization between two individuals leads to a polyclonal antibody response in which many different antibodies are elicited. The antiserum is composed of antibodies each specific for a particular antigen. The serum can be tested against a panel (random sample) of individuals to examine the distribution of those antigens recognized by that serum. Because the antiserum is a heterogeneous mixture of antibodies, sometimes it is first necessary to reduce the complexity by adsorption or by adsorption and elution. Adsorption consists of removing unwanted antibodies by reacting the antiserum with a cell. If the cell expresses antigens reactive with some of the antibodies in the serum, these antibodies will bind to the cells. The cells together with the bound antibody can then be removed. This leaves only the unreactive antibodies left in the serum. Alternatively, if an antibody of a particular specificity is wanted, it can be adsorbed on to the antigen. Antigen–antibody complexes can be separated from the non-reactive antibodies. The complex can be dissociated and the positively selected antibody eluted.

One of the best ways to reduce the heterogeneity of the antisera is to use inbred and/or congenic mice. Animals of one inbred strain are immunized with cells from a congenic strain. The resulting antiserum is tested against cells from all available mouse strains. If the antiserum reacts with cells from many strains, adsorption is performed. For example, if B10 mice are immunized with B10.A cells, a serum reactive with B10.A, B10.K and B10.D2 cells is produced (Table 1.2). If this serum is adsorbed with B10.K cells, reactivity remains for B10.D2 and B10.A cells. If the serum were adsorbed with B10.D2 cells instead of B10.K, reactivity would remain for B10.A and B10.K cells. As expected, adsorption with B10.A cells completely removes all activity not only to B10.A cells but also to B10.K and B10.D2 cells. B10 cell adsorption has no effect (Table 1.3) as only antigens present on B10.A were used to elicit the immune response. If the B10 anti-(B10.A) serum was first adsorbed with B10.K

Table 1.2 Primary adsorption analysis of a typical alloantiserum: adsorption of B10 anti-(B10.A) serum

	Serum adsorbed with				
Test cell	None	B10.A	B10.K	B10.D2	B10
B10.A	+	0	+	+	+
B10.K	+	0	0	+	+
B10.D2	+	0	+	0	+
B10	0	0	0	0	0

Table 1.3 Secondary adsorption analysis of a typical alloantiserum: adsorption of B10 anti-(B10.A [B10.K]) serum

	Serum adsorbed with				
Test cell	None	B10.A	B10.K	B10.D2	B10
B10.A	+	0	+	0	+
B10.K	0	0	0	0	0
B10.D2	+	0	+	0	+
B10	0	0	0	0	0

(designated B10 and B10.A [B10.K]) and then adsorbed with B10.D2 before testing on B10.A cells, further insight into the relationship between B10.A, B10.K and B10.D2 could be obtained. Here we see that adsorption with B10.D2 not only removed activity for itself but also for B10.A. This serum {B10 anti-(B10.A [B10.K])} is now operationally monospecific, since we cannot define any more specificities, given the strains available. We know from this experiment that antigens present on B10.K cells are also present on B10.A but not B10.D2, while antigens present on B10.D2 are present on B10.A but not B10.K cells. Further, all specificities found on B10.A are present either on B10.K or B10.D2. We can now define two specificities, and four phenotypes. B10.A has specificities 1 and 2. B10.K expresses only specificity 1, and B10.D2 specificity 2. B10 is a null phenotype with respect to these specificities.

This method of adsorption analysis defines antigens in different strains. Classical genetic segregation analysis allows the determination of the relative genetic locations of the genes encoding these antigens. Unfortunately, genetic analysis tells us nothing about the biochemical or physical properties of a given antigen or its relationship to other antigens. Obviously, a detailed serological analysis depends upon the knowledge of the mouse strains used. Therefore, antigens or differences between strains can only be defined within the limits of the number of inbred strains available for study.

Another potential source for heterogeneity in a serological reaction lies in the antibodies and antigens themselves. For instance, although antibodies possess unique antigen-combining sites, they often cross-react with antigens which either share identical antigenic determinants or determinants which are stereochemically similar. A given antigen may possess more than one antigenic determinant per molecule and thus provoke the response of several distinct antibody families. Therefore, although each single antibody is derived from an individual B cell clone and is specific for one antigenic determinant, it is very difficult to produce a truly monospecific antiserum by a classical immunization protocol. This problem can be circumvented by a technique developed by

Kohler and Milstein (1975, 1976) which results in the production of monoclonal antibodies.

1.5 HYBRIDOMAS AND MONOCLONAL ANTIBODIES

Antibodies arise from clonal populations of B cells which produce antibodies each with unique antigen-combining sites. Each B cell carries a membrane-bound antibody on its cell surface which, when triggered by antigen, causes the B cell to undergo a process of clonal expansion in which the B cell is stimulated to divide extensively. Most of the B cells then mature and secrete antibodies. These mature antibody-secreting cells are called plasma cells. The remaining B cells are reserved in a state of readiness to respond to antigen if it is encountered again and are called memory cells. Although each antibody arises from a clonal population of B cells, most responses are polyclonal. In most immunizations the immunogen is a mixture of different antigens each with multiple determinants. Kohler and Milstein (1976) were able to fuse a B cell with a myeloma cell to establish a cell line (hybridoma) which produces a monoclonal antibody which is specific for a single antigenic determinant.

Normally activated B cells which mature and secrete antibody do not divide. Therefore they cannot be propagated indefinitely in tissue culture. However, myeloma cells which are plasma cell tumours have infinite lifespans and are easily propagated in culture. The two cell types are fused by an agent such as polyethylene glycol or Sendai virus. After fusion both B cell myeloma hybrids and the unfused myeloma cells can potentially survive and propagate in culture. However, since the myeloma cells are selected to be defective in the purine salvage pathway, the addition of hypoxanthine, aminopterin and thymidine (HAT) to the growth medium kills them (see HAT selection, Chapter 3). The resultant hybridoma line retains the antigen specificity of the B cell and the immortality of the myeloma cell. The hybrids can be propagated in tissue culture or passaged in a syngeneic (genetically identical) mouse as an ascites tumour. Ascites tumours grow as single cell suspensions in the peritoneal cavity. Peritoneal ascites fluid containing the monoclonal antibodies secreted by the tumour can be harvested from the peritoneal cavity.

The hybridomas are screened by collecting culture fluid supernatants or ascites which contain monoclonal antibodies and testing them for binding activity to the antigen. Although a monoclonal antibody does not define physical or biochemical properties of an antigen or its relationship to other antigens better than conventional antisera, it does arise from a single B cell and thus has a unique antigen combining site to one antigenic determinant. Thus, it will bind to only a single determinant on an antigen. For instance, when testing several monoclonal antibodies which arise from an immunization with a pure protein antigen, one may find that adsorption to the antigen with one monoclonal antibody blocks reactivity with another monoclonal antibody. This indicates that both monoclonals react with the same determinant or determinants which are close stereochemically. Often binding with one monoclonal antibody does not block reactivity with another. This suggests that each monoclonal antibody recognizes a unique, spatially separate determinant. Hence, monoclonals are able to define domains of individual molecules as well as differentiating between separate molecules.

1.6 β_2-MICROGLOBULIN

β_2-Microglobulin is a non-glycosylated 12000 dalton protein which is non-covalently associated with several chromosome 17 encoded antigens (Nilsson *et al.*, 1973; Poulik, 1973; Fanger and Bernier, 1973; Ostberg *et al.*, 1975; Vitetta *et al.*, 1975). Although associated with integral membrane proteins, it is not membrane bound (Steck, 1974). β_2-Microglobulin bears a striking amino acid and nucleic acid sequence homology to one of the constant-region domains of immunoglobulins (Smithies and Poulik, 1972). It was therefore interesting to investigate the genetics of β_2m in mice. However, because of the absence of suitable serological and genetic reagents, only recently have the genetic investigations of β_2m advanced. These advances are due to the discovery of a β_2m polymorphism among inbred mouse strains. In addition, congenic mice as well as recombinant inbred strains have been used to localize the β_2m gene.

β_2-Microglobulin was first isolated in 1968 from the urine of humans with Wilson's disease, a renal tubular dysfunction (Berggard and Bearn, 1968). Subsequently, it has also been found free in various human body fluids including serum, amniotic fluid, milk and seminal fluid (Evrin and Wibell, 1972; Evrin *et al.*, 1971). The striking resemblance of β_2m to the constant domains of immunoglobulins as well as the demonstration that β_2m was present on the lymphocyte surface prompted the suggestion that β_2m might represent a free immunoglobulin domain. However, demonstrations that β_2m was synthesized not only by lymphocytes, but also by a variety of cell types, such as mesenchymal and epithelial cells, indicated that it plays a more general role in cell surface biology (Nilsson *et al.*, 1973; Hutteroth *et al.*, 1973; Bernier and Fanger, 1972). These data together prompted the suggestion that both immunoglobulin and β_2m molecules may have evolved from a common ancestor but that β_2m was not functionally directly related to immunoglobulins. It was speculated that the immune system may represent a highly specialized version of a more general cell recognition system.

A biochemical association between β_2m and HLA antigens was first postulated by Nakamuro *et al.* (1973). They found that an 11000-dalton peptide isolated with HLA antigens had an amino acid composition identical with urinary β_2m. Subsequently Poulik *et al.* (1973) demonstrated the close proximity of HLA and β_2m on the cell surface by co-capping experiments. Since cell surface molecules are mobile, they can be cross-linked by specific reagents to form patches on the cell membrane. These patches of cross-linked molecules move toward the pole of the cell and coalesce into a polar cap which is eventually ingested or shed by the cell. If two molecules are associated on the cell membrane, capping of one molecule with a divalent antibody will result in pulling the other molecule along into the cap. This is called co-capping. Poulik *et al.* showed that anti-(β_2m) serum was able to co-cap the HLA determinants on the surface of human lymphocytes. Furthermore, it was found that coating lymphocytes *in vitro* with some anti-(β_2m) serum could block binding of HLA antibodies. This implied a close association between β_2m and HLA on the cell surface. While capping and blocking studies were underway, several investigators (Grey *et al.*, 1973; Cresswell *et al.*, 1973, 1974; Peterson *et al.*, 1974) used immunochemical techniques to demonstrate the molecular association between β_2m and HLA class I antigens.

Following these studies, β_2m or an analogous 12000 dalton polypeptide was found non-covalently bound to antigens of the major histocompatibility complex of all species investigated. In addition, a 12000 dalton polypeptide has also been found to be non-covalently associated with Tla, a thymocyte differentiation antigen. Tla is a 44000 dalton glycoprotein (Stanton *et al.*, 1975; Vitetta *et al.*, 1976) which is found exclusively on thymocytes and is coded for by a gene closely linked to the MHC on chromosome 17 of the mouse.

Vitetta *et al.* (1976) investigated the identity of the β_2m associated with Tla and H-2 with a cross-reacting anti-(rat β_2m) serum. Mouse thymus cells were surface-labelled with ^{125}I, followed by detergent extraction. To determine the structure of the anti-(β_2m)-reactive proteins, the precipitates were separated by electrophoresis by two methods. First the samples were reduced to dissociate disulphide-linked chains and separated on sodium dodecyl sulphate (SDS)/polyacrylamide gels. Alternatively, SDS/polyacrylamide gel electrophoresis was performed under non-reducing conditions, which dissociate non-covalent but not disulphide bonds. This gives the molecular weight of the molecule with disulphide bonds intact. Assuming that the specificity of the antiserum is solely to β_2m several conclusions were drawn:

(1) All molecules which react with this serum have subunits of molecular weight 44000 and 12000.

(2) The major subunits recovered from the cell lysate appear to be non-covalently bound to β_2m (i.e. they are dissociated from β_2m under reducing conditions).

(3) All of the immunoprecipitated β_2m is bound to either H-2 or Tla molecules.

Vitetta *et al.* found that immunoprecipitation with antibodies against the 44000 dalton component also removed β_2m. These studies support earlier evidence that H-2 molecules are associated with β_2m. Furthermore, the 12000 dalton protein associated with Tla is β_2m.

The structural relationship between β_2m and immunoglobulin constant domains coupled with an association between β_2m and MHC antigens on the cell surface raised the question of the genetic location of the β_2m genes. In 1975 two laboratories (Goodfellow *et al.*, 1975; Faber *et al.*, 1976) using synteny mapping in human–mouse somatic cell hybrids showed the gene for β_2m mapped to human chromosome 15 and not chromosome 6 where HLA is located (see Chapters 3 and 4). However, until recently, mapping of the mouse β_2m gene by conventional segregation tests had been precluded by the lack of any detectable β_2m polymorphism. Michaelson *et al.* (1980) reported a structural variation in mouse β_2m. Cells from mouse spleen or thymus were first labelled with ^{125}I or [^{35}S]methionine, solubilized with NP-40, and then immunoprecipitated with anti-(β_2m) serum. The β_2m from mouse strains C57Bl/6, B10.A and B6-Tla^a migrate slightly faster in SDS/polyacrylamide gels than β_2m from mouse strain A. Both β_2m bands are seen in (B6×A)F$_1$ mice which suggests that β_2m is expressed co-dominantly (Fig. 1.3).

Michaelson designated the gene coding for β_2m, *B2M,* and suggested that the β_2m variants were probably due to variation in the structural gene rather than post-transcriptional modification, since most modifiers act dominantly. The results of Michaelson *et al.* also suggested that, as in man, the *B2M* locus is not closely linked to the MHC. Although B6 and B10.A possessed different *H-2* alleles they have the same *B2M* allele, while B10.A and A possess the same *H-2* but have different *B2M* types (Table 1.4). If *B2M* was closely linked to the

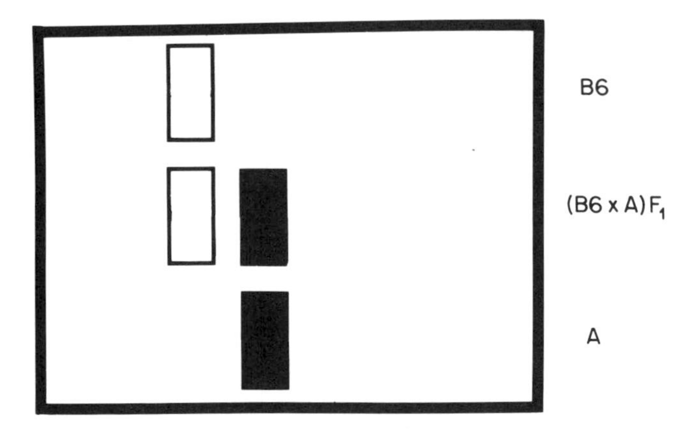

Fig. 1.3 Schematic representation of SDS/polyacrylamide gel electrophoresis of ^{35}S-labelled material from B6, (B6×A)F$_1$ and A strain cells precipitated by anti-(β_2m) serum demonstrating co-dominant expression of β_2m variants. Black = *B2M*^{b}; white = *B2M*^{a}.

Table 1.4 *H-2* and *B2M* types of congenic mouse strains used by Michaelson *et al.* (1980) to detect β_2m polymorphism

	H-2		*B2M*
	K	*D*	
B6	b	b	a
B10.A	k	d	a
B6-Tla^a	b	b	a
A	k	d	b

MHC, mice which shared the same MHC (as in B10.A and A) would possess the same *B2M* allele.

In a further investigation of β_2m polymorphism, Robinson *et al.* (1981) surveyed the *B2M* alleles of a variety of strains with different *H-2* haplotypes. Similarly, these investigators found two forms of β_2m which they called *B2M*^{f} and *B2M*^{s} for faster and slower migrating forms on polyacrylamide gel electrophoresis. Furthermore Robinson *et al.* confirmed that both alleles are expressed co-dominantly in heterozygotes.

Shortly after his initial description of β_2m polymorphism by polyacrylamide gel electrophoresis, Michaelson (1981) mapped the *B2M* gene to mouse chromosome 2, using a series of recombinant inbred (RI) strains derived from a C57BL/6×Balb/c cross. Michaelson examined the *B2M* alleles of seven C×B strains and noted that the *B2M* had an identical strain distribution pattern to the minor transplantation antigen gene *H-3* (Table 1.5) which implied linkage of *B2M* and the chromosome 2 marker *H-3*. Linkage was confirmed by an analysis of β_2m from two strains of *H-3* congenics (mice differing at *H-3*, identical otherwise) which showed that the *B2M* allele was characteristic of the *H-3* donor, rather than the background strain. In addition, β_2m from a number of inbred strains was analysed by SDS/polyacrylamide gel electrophoresis. To date, all *H-3*^{a} strains are *B2M*^{b} while all non-*H-3*^{a} strains have been *B2M*^{a}. Michaelson (1981) points out that while identity would imply that the *H-3* molecule is itself β_2m and that *H-3* transplantation antigenicity is due to β_2m structural polymorphism, we do not know whether the histogenetically defined *H-3* polymorphism and *B2M* variation are single locus or two closely linked loci.

The *H-3* region also codes for a cell surface alloantigen Ly-4 (Snell *et al.*, 1973; McKenzie and Snell, 1975; McKenzie, 1975) Ly-4 has been postulated to be a serological determinant on *H-3*. *H-3* and *Ly-4* have not been separated by

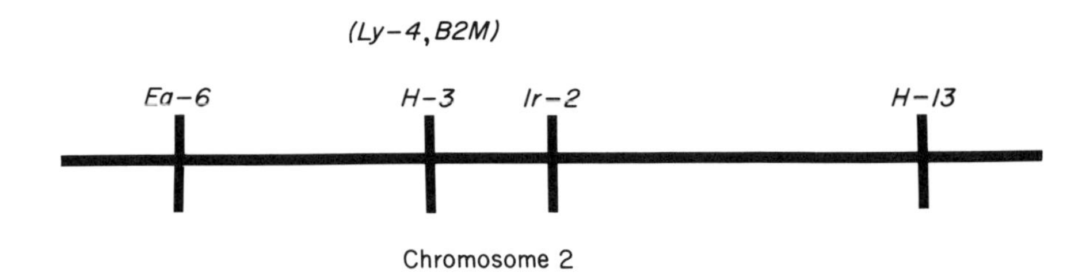

Fig. 1.4 Map of region of mouse chromosome 2 to which β_2m maps.

recombination (McKenzie and Potter, 1979). Hence, *H-3, Ly-4* and *B2M* could be closely linked but separate genes or alternatively a single gene with pleiotropic effects. Furthermore, *H-3* is linked to loci controlling immune responses *Ir-2* (Gasser, 1969; Gasser and Shreffler, 1974), a locus coding for erythrocyte antigen *Ea-6* (Lilly, 1967) and another minor histocompatibility gene *H-13* (Graff *et al.*, 1978) (Fig. 1.4).

Table 1.5 H-2, H-3 and β_2m phenotypes of Balb, B6 and C×B recombinant inbred strains used by Michaelson (1981) to map β_2m to chromosome 2

	H-2	H-3	B2M
Balb	b	c	a
B6	d	a	b
C×BD	d	c	a
C×BH	b	c	a
C×BE	b	a	b
C×BG	d	a	b
B×BI	b	a	b
C×BJ	b	a	b
C×BK	b	a	b

Cox *et al.* used a panel of chinese hamster–mouse somatic cell hybrids to confirm the assignment of the *B2M* gene to chromosome 2. These hybrid cells discard mouse chromosomes. Karyotyping coupled with analysis of selected mouse enzyme markers provided data on the segregation of chromosome 2 and the *B2M* gene. Cox *et al.* (1982) correctly points out that it is unlikely that *Ly-4* and *B2M* are identical since Ly-4 is expressed only on lymphocytes while β_2m is expressed on most nucleated cells.

The early studies on β_2m using anti-(β_2m) and anti-(MHC) sera in combination with immunochemical separation techniques were able to provide substantial evidence of the association of β_2m with major histocompatibility antigens. Although human β_2m was mapped to chromosome 15 by the use of somatic cell hybrids, it was not until Michaelson *et al.* used immunoprecipitation techniques that β_2m polymorphism was discovered in mice. Thereafter, conventional mapping by linkage analysis could proceed. The availability of recombinant inbred strains, in this case the C×B series, allowed Michaelson to associate the distribution of *B2M* alleles with those of a polymorphic locus on mouse chromosome 2, *H-3*. Hence, the *B2M* locus has been mapped to the *H-3* region of chromosome 2.

The association of β_2m with MHC determinants revealed the distribution of β_2m on cell membranes. The use of an immunogenetic approach has pinpointed the chromosomal location of *B2M*. This is important because the structural resemblance of β_2m to immunoglobulin constant domains and its association with MHC antigens has raised many questions about the possible genetic relationship of these molecules and their gene families. Furthermore, the discovery of β_2m polymorphism and its association with already polymorphic H-2 antigens raises further immunological questions about the function of their association on the cell surface. While we are still largely uncertain of the function of β_2m on the surface of most cells, its study demonstrates the interdependence of classical serology and the use of congenic mouse strains in characterizing cell surface molecules.

1.7 LY-5/T200

The various facets of the immune response are performed by the two major lymphocyte cell types: T cells, which mediate delayed type hypersensitivity and regulate antibody formation, and B cells, which produce immunoglobulin molecules. Because each cell type has a separate, characteristic role in the immune response, it is important to differentiate between them. In the early 1960s Reif and Allen (1963) described the Thy-1 antigen as a differentiation antigen of thymus and brain. Other investigators studied a variety of cell surface antigens which were expressed preferentially on T or B cells. The best-known examples are Thy-1, extensively studied by Raff (1971), and the Ly-1,2,3 markers first described by Boyse and Old (1969) and extensively studied by Cantor and Boyse (1975). The Ly-1,2,3 markers are differentially expressed on T cell subsets.

In 1975 two groups, Trowbridge *et al.* (1975) and Komuro *et al.* (1975), were searching for T and B cell markers. They both defined a glycoprotein with a molecular weight of approximately 200 000. Komuro found the protein through a programme of alloimmunizations designed to define new T cell *Ly* loci and called the polymorphic protein Ly-5. Trowbridge utilized rabbit anti-(mouse lymphocyte) serum and immunoprecipitation to isolate the protein found on T cells. He later named this molecule T (for T cell) 200 (for its molecular weight).

At the time *Ly-5* was described, four genetic loci of the mouse that appeared to contribute specifically to the constitution of the T lymphocyte surface had been described: *Thy-1* (Raff, 1971), *Tla, Ly-1* (Boyse and Old, 1969) and *Ly-2/Ly-3* (Itakura *et al.*, 1972). Lymphocyte antigens were studied with the hope of finding a precise correlation of antigen expression and cell functions. The Ly-5 antigens were found immunizing recipients with donor tissue compatible for H-2 and other known T cell alloantigens. These donor–recipient combinations were designed to produce antibodies directed at the products of previously unknown *Ly* loci. Anti-(Ly-5) serum was first produced in SJL mice immunized with cells from A.SW mice and (B6-H-2^k × A.SW)F_1 mice immunized with SJL cells. Segregation tests revealed that each antiserum recognized a predominantly single genetic factor. The strain distribution of Ly-5 antigens suggested that the two anti-(Ly-5) sera defined two alleles. Each serum recognized an antithetical alloantigen Ly-5.1 and Ly-5.2. Every strain tested expressed either one or the other antigen. Segregation tests in F_2 mice suggested that the two antigens were coded by alleles (Table 1.6). Linkage tests revealed no linkage to the tested markers.

Michaelson *et al.* (1979) used thymocyte or spleen cell extracts precipitated with either anti-(Ly-5.1) or anti-(Ly-5.2) sera. When the molecules were separated by SDS/polyacrylamide gel electrophoresis, two major proteins of 175 000 and 220 000 daltons were observed. With thymocyte extracts, a single molecular species of 175 000 daltons was seen; however, when spleen cell extracts were precipitated, three size classes of 175 000, 185 000 and 220 000 daltons were observed. Both thymocyte and spleen antigens were controlled by the *Ly-5* locus or genes within the vicinity. Anti-(Ly-5.1) serum precipitated material from B6 (Ly-5.1) but not from the B6-Ly-5^b. Antithetical results were

Table 1.6 Immunization and segregation data for production of Ly-5 antisera specifying the antigens Ly-5.1 and Ly-5.2

Serum	Specificity	Classes of progeny				
SJL anti-(A.SW)	Ly-5.1	+	+	−	−	
(B6-H-2^k × A.SW)F_1 Anti-(SJL)	Ly-5.2	+	−	+	−	
Number observed in each class:		35	15	16	0	
Expected for two loci:		37.1	12.3	12.3	4.1	$\chi^2 = 5.97, P < 0.05$
Expected for one locus:		35	16.5	16.5	0	$\chi^2 = 0.15, P > 0.5$

obtained with anti-(Ly-5.2) serum. Further, SJL mice (Ly-5.2) yielded a gel pattern identical with B6-Ly-5^b. Balb/c (Ly-5.1) yielded a pattern identical with B6 (Ly-5.1). A rather surprising result was that the spleen extracts of athymic mice (Balb-nu/nu), which lack T cells, expressed only the 220000 dalton protein. The 220000 dalton material precipitated by Ly-5 antisera must therefore be expressed on cells other than T cells.

Conclusive proof that Ly-5 was expressed by non-T cells was obtained by Scheid and Triglia (1979). They further characterized the cells which expressed Ly-5 by using the Ly-5 congenic mouse strain (B6-Ly-5^b) in combination with the protein A–sheep red blood cell rosette assay and cytotoxicity tests. In the cytotoxicity assay, cells are incubated with test antisera at various dilutions and rabbit complement is added subsequently. The antigen-positive cells are lysed. In the PA–SRBC assay *Staphylococcus aureus* protein A (PA), an immunoglobulin (Ig) binding protein, is coupled to sheep red blood cells (SRBC). Test cells, previously incubated with antiserum are added to the PA–SRBC. The antigen-positive cells which had adsorbed antibodies formed rosettes with the *Staph. A*-coated SRBCs. The *Staph. A* bound to the SRBC acting as a bridge also binds the Fc portion of the test antibodies. This assay was used in addition to the cytotoxicity assay because it can be more sensitive; often it could detect Ly-5-positive cells not lysed in the cytotoxicity assay. Scheid and Triglia found that most cells from haematopoietic lineage, including B cells, macrophages and natural killer cells, expressed Ly-5. A notable serological feature of Ly-5 expression was the sensitivity of T cells relative to other cells in the cytotoxic assays. They suggested that the disparity between the two systems might lie in the number of the Ly-5 molecules present on the cell surface.

Trowbridge *et al.* (1975) also began probing the lymphocyte membranes of both T and B cells in the hope of identifying molecules which were specific for each cell type. The approach taken was to characterize biochemically the surface membranes of pure populations of B and T cells, obtained by selectively stimulating spleen cells and thymocytes, with either T or B cell mitogens. Lipopolysaccharide (LPS) was used to selectively stimulate B cells. Either concanavalin A (ConA), pea lectin or allogeneic cells were used to stimulate T cells. T or B cell enrichment was checked by monitoring ^{51}Cr release as a measure of T cell lysis induced by anti-(Thy-1) serum and complement (Fig. 1.5).

The rationale behind this approach is described. Following labelling with ^{51}Cr, the T cells are washed to remove any ^{51}Cr not taken up by the cells. Antiserum and complement are added to the ^{51}Cr-labelled cells. T cells that express the antigen recognized by the antibody are lysed. The ^{51}Cr released by the killed cells is a measure of the antigen–antibody reaction. B cells, which should not have been stimulated by T cell mitogens or allogenic cells, should not be lysed by anti-(Thy-1) serum which is specific for T cells. Thus B-enriched populations do not release ^{51}Cr following treatment with anti-(Thy-1) serum

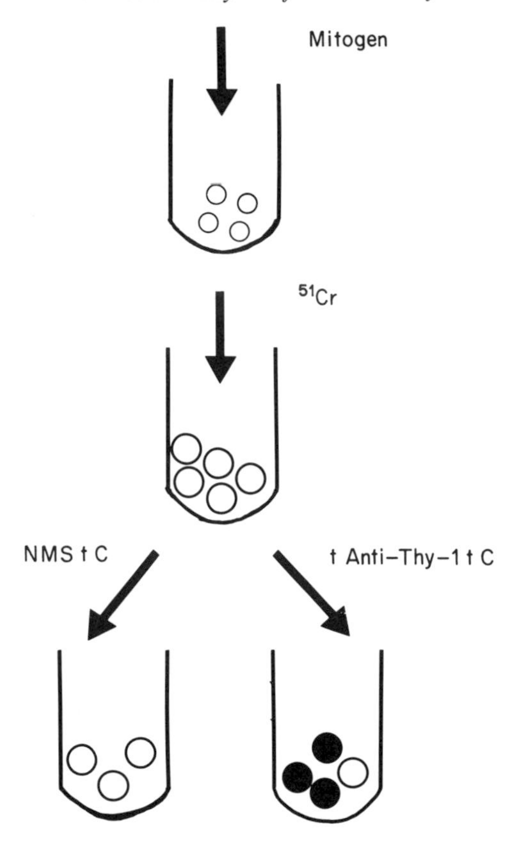

Fig. 1.5 Schematic representation of ^{51}Cr release assay.

and complement. When T or B cells were selectively stimulated, over 90% of ConA-stimulated cells and less than 2% of LPS-stimulated cells were lysed with anti-(Thy-1) serum and complement.

Selectively stimulated populations of T or B cells were cell surface-radiolabelled with ^{125}I by lactoperoxidase. Following extraction of the cell, the surface molecules were examined by SDS/polyacrylamide gel electrophoresis. Proteins from labelled thymocytes, T cell lymphomas and both normal and activated peripheral T cells gave a broad band of radioactivity from 170000 to 190000 daltons. This band was absent from gels run with B cell-extracted material which instead ran in a narrow molecular-weight range near 220000. Trowbridge suggested these iodinated proteins represented marker proteins specific for T or B cells. Furthermore, both the 220000 and 170000 dalton proteins could be precipitated by rabbit anti-(mouse lymphocyte) serum. This result suggested that these proteins must be the major cell surface antigens recognized by rabbit anti-(mouse lymphocyte) antibodies.

Subsequently, Trowbridge and colleagues found another T cell protein of 25 000 daltons which was recognized by rabbit anti-mouse antibodies. For this reason, T cell proteins were given the designations T25, for 25 000 dalton T cell protein and T200 for the 200 000 dalton protein described earlier.

Further characterization of the T200 and T25 glycoproteins revealed a peculiar spectrum of properties surrounding T200; paradoxical results were often seen. Although B cells could absorb the anti-T200 activity of anti-lymphocyte sera, no T200 protein products could be detected in B cell immunoprecipitates. Similarly, antisera raised against T lymphocytes precipitated a 170 000 dalton protein from T lymphocytes, but not a 220 000 dalton protein from B lymphocytes. Hence, it was suggested that T200 and the B cell protein might share common antigenic determinants and be structurally related. The differences in the determinants, perhaps in the carbohydrate moieties, could play a role in determining the anatomical distribution of B and T cells in lymphoid tissues (Trowbridge and Mazauskas, 1976).

Trowbridge (1978) described the production of a hybridoma secreting monoclonal, I3/2, an anti-T200 antibody. Immunoprecipitation of ^{125}I-labelled extracts from thymocytes, normal spleen cells or separated T and B cells showed that I3/2 precipitates a 220 000 dalton molecule from B cells as well as a 190 000 dalton protein from T cells. Hence, the studies of the specificity of the monoclonal antibody produced by the rat spleen–mouse myeloma hybrid (I3/2) established that a single antibody reacts with both the high molecular weight antigens of T and B lymphocytes. Subsequent to the development of a monoclonal antibody specific for T200, Trowbridge's group noticed that the tissue distribution and biochemical properties of T200 were very similar to those of the Ly-5. To investigate the relationship between T200 and Ly-5, Omary *et al.* (1980), used a T200-negative mutant cell line [BW5147 (T200$^-$)] which does not express T200. BW5147 (T200$^-$) is derived from an AKR thymoma, BW5147. AKR mice express Ly-5.1. Hence, if Ly-5.1 were expressed on the BW5147 but not on T200 cells, this would provide strong evidence that anti-(Ly-5) sera detect a polymorphism of the T200 molecule. The presence of the Ly-5 alloantigen was tested by both antibody binding and immunoprecipitation. While anti-(Ly-5.1) serum bound to BW5147 cells, anti-(Ly-5.1) serum bound to T200$^-$ mutant cells only slightly more than the Ly-5.2-negative control serum. Monoclonal T200 bound only to BW5147 cells.

Immunoprecipitation studies showed that both Ly-5.1 antisera and monoclonal T200 antibody precipitated molecules from BW5147 with apparent molecular weights of 200 000 that co-migrated on SDS/polyacrylamide gels. Neither antibody precipitated a labelled species from T200$^-$ cells. In addition, tryptic peptide maps of T200 and Ly-5.1 molecules appeared identical. These results established that Ly-5 alloantisera indeed detected a polymorphism of T200.

Previous estimates of T200 cell surface antigen density utilized monoclonal anti-T200 in a saturation binding assay and flow cytometry analysis. These studies showed approximately 50000 molecules of T200 on T cells and approximately 25000 on B cells (Morris and Williams, 1975). Thus, Ly-5 is similar in concentration on the cell surface to other alloantigens of murine lymphocytes, such as Thy-1, Ly-1 and Ly-2. The Ly-5 alleles are a polymorphism of an abundant cell surface glycoprotein.

Siadak and Nowinski (1980) confirmed and extended the results of Omary *et al.* by utilizing immunoabsorbent columns made by covalently linking anti-T200 monoclonal antibody to Sepharose 4B. ^{125}I-surface-labelled membrane preparations containing Ly-5 and T200 antigens were passed through the immunoabsorbent column. Prior to passage through the column, the membrane preparation contained antigenic material reactive with both anti-(Ly-5) and anti-T200 sera. Both activities were removed after passage through the anti-T200 column. This procedure did not result in depletion of either H-2 or β_2m which were used as negative controls. Therefore, the results of Siadak and Nowinski coincide with those of Trowbridge's group. Both groups agree that Ly-5 alloantisera detect a polymorphism of T200.

In 1975, both Trowbridge's and Komura's groups, searching for new T and B cell markers, inadvertently described the same cell surface molecule. Trowbridge and colleagues used conventional rabbit anti-(mouse lymphocyte) sera and immunoprecipitation to define a major cell surface glycoprotein, T200, found on haemopoietic cells. Komura and colleagues immunized inbred mice compatible for all known H-2 and T cell alloantigens. The serum produced by the immunizations defined Ly-5 which was originally described as an exclusive T cell marker. Actually it recognizes a polymorphism of the T200 molecule.

It was the development of an anti-T200 monoclonal and the subsequent selection of a T200$^-$ mutant cell line using the T200 monoclonal that finally allowed Omary *et al.* to link the two molecules. These data showed directly that anti-(Ly-5) sera defines a polymorphism of T200. Thus the study of Ly-5/T200 demonstrates the importance of the development of monoclonals in distinguishing between multiple determinants expressed on one molecule versus cross-reactive determinants expressed on different molecules.

1.8 CODA

The β_2m and Ly-5/T200 systems demonstrate how genetically inbred mouse strains, used in conjunction with careful serology, can be used to characterize the molecules associated with particular cell surface phenomena.

Recent advances in molecular biology have allowed MHC genes to be cloned, isolated and introduced into mouse L cells by DNA-mediated gene transfer (Goodenow *et al.*, 1982; and Chapter 4). Until now identification of the gene products as well as their expression on the cell surface has been characterized primarily by the use of monoclonal antibodies to specific genetic regions defined by combinations of congenic strains.

Now the prospect exists of custom-designing genetic changes. New 'designer genes' can be introduced into cells or embryos to probe the structure and function of cell surface molecules.

REFERENCES

Bailey, D.W. (1971), *Transplantation,* **11,** 325–327.

Benacerraf, B. and McDevitt, H.O. (1972), *Science,* **175,** 273–279.

Berggard, I. and Bearn, A.G. (1968), *J. Biol. Chem.,* **243,** 4095–4103.

Bernier, G.M. and Fanger, M.W. (1972), *J. Immunol.,* **109,** 407.

Boyse, E.A. and Old, L.J. (1969), *Annu. Rev. Genet.,* **3,** 269–290.

Cantor, H. and Boyse, E.A. (1975), *J. Exp. Med.,* **141,** 1377–1389.

Cox, D.R., Sawaki, J.A., Yee, D., Appela, E. and Epstein, C.J. (1982), *Proc. Natl. Acad. Sci. U.S.A.,* **79,** 1930–1934.

Cresswell, P., Turner, J.M. and Strominger, J.L. (1973), *Proc. Natl. Acad. Sci. U.S.A.,* **70,** 1603–1607.

Cresswell, P., Springer, T., Strominger, J.L., Turner, M.J., Grey, H.M. and Kubo, R.T. (1974), *Proc. Natl. Acad. Sci. U.S.A.,* **71,** 2123–2127.

Evrin, P.E. and Wibell, L. (1972) *Scand. J. Clin. Lab. Invest.,* **29,** 69–74.

Evrin, P.E., Peterson, P.A., Wide, L. and Berggard, I. (1971), *Scand. J. Clin. Lab. Invest.,* **28,** 439–443.

Faber, M.F., Kucherlapati, R.S., Poulik, M.D., Ruddle, F.H. and Smithies, O. (1976), *Somatic Cell Genet.,* **2,** 141–157.

Fanger, M.W. and Bernier, G.M. (1973), *J. Immunol.,* **111,** 609–617.

Gasser, D.L. (1969), *J. Immunol.,* **103,** 66–70.

Gasser, D.L. and Shreffler, D.C. (1974), *Immunogenetics* **1,** 133–140.

Goodenow, R., McMillan, M., Orn, A., Nicolson, M., Davidson, N., Frelinger, J.A. and Hood, L. (1982), *Science,* **215,** 677–679.

Goodfellow, P.N., Jones, E.A., van Heyningen, V., Solomon, E., Bobrow, M., Miggiano, V. and Bodmer, W.F. (1975), *Nature (London),* **254,** 267–269.

Gorer, P.A., Lyman, S. and Snell, G.D. (1948), *Proc. R. Soc. London Ser. B,* **135,** 499–505.

Graff, R.T., Brown, D.H. and Snell, G.D. (1978), *Immunogenetics,* **7,** 413–423.

Green, E.L. (1966), in *Biology of the Laboratory Mouse* (E.L. Green, ed.), McGraw-Hill, New York, pp. 11–22.

Green, E.L. and Doolittle, D.P. (1963), in *Methodology in Mammalian Genetics* (W.J. Burdette, ed.), Holder-Day, San Francisco, pp. 3–55.

Grey, H.M., Kubo, R.T., Colon, S.M., Poulik, M.D., Creswell, P., Springer, T., Turner, M. and Strominger, J.L. (1973), *J. Exp. Med.,* **138,** 1608–1612.

Hutteroth, T.H., Cleve, H., Litwin, S.D. and Poulik, M.D. (1973), *J. Exp. Med.,* **137,** 838–843.

Itakura, K., Hutton, J.J., Boyse, E.A. and Old, L.J. (1972), *Transplantation,* **13,** 239–243.
Kohler, G. and Milstein, C. (1975), *Nature (London),* **265,** 495–497.
Kohler, G. and Milstein, C. (1976), *Eur. J. Immunol.,* **6,** 511–519.
Komuro, K., Itakura, K., Boyse, E.A. and John, M. (1975), *Immunogenetics,* **1,** 452–456.
Lilly, F. (1967), *Transplantation,* **5,** 83–84.
McKenzie, I.F.C. (1975), *J. Immunol.,* **114,** 856–862.
McKenzie, I.F.C. and Potter, T. (1979), *Adv. Immunol.,* **27,** 181–338.
McKenzie, I.F.C. and Snell, G.D. (1975), *J. Immunol.,* **114,** 848–855.
Michaelson, J. (1981), *Immunogenetics,* **13,** 167–171.
Michaelson, J., Scheid, M. and Boyse, E.A. (1979), *Immunogenetics,* **9,** 193–197.
Michaelson, J., Rothenberg, E. and Boyse, E.A. (1980), *Immunogenetics,* **11,** 93–95.
Morris, R.J. and Williams, A.F. (1975), *Eur. J. Immunol.,* **5,** 274–285.
Nakamuro, K., Tanigaki, N. and Pressman, D. (1973), *Proc. Natl. Acad. Sci. U.S.A.,* **70,** 2863–2865.
Nilsson, K., Evrin, P.E., Berggard, I. and Ponten, J. (1973), *Nature (London) New Biol.,* **244,** 44–45.
Omary, M.B., Trowbridge, I.S., and Scheid, M.P. (1980), *J. Exp. Med.,* **151,** 1311–1315.
Ostberg, L., Rask, L., Wigzell, H. and Peterson, P.A. (1975), *Nature (London),* **253,** 735–737.
Peterson, P.A., Cunningham, B.A., Berggard, I. and Edelman, G.M. (1972), *Proc. Natl. Acad. Sci. U.S.A.,* **69,** 1697–1701.
Peterson, P.A., Rask, L. and Lindblom, J.B. (1974), *Proc. Natl. Acad. Sci. U.S.A.,* **71,** 35–39.
Poulik, M.D. (1973), *Immunol. Commun.,* **2,** 403.
Poulik, M.D., Bernoco, M., Bernoco, D. and Ceppellini, R. (1973), *Science,* **182,** 1352–1354.
Poulik, M.D., Ferrone, S., Pelligrino, M.A., Sevier, D.E., Oh, S.K. and Reisfield, R.A. (1974), *Transplant. Rev.,* **21,** 106.
Raff, M.C. (1971), *Transplant. Rev.,* **6,** 52–80.
Reif, A.E. and Allen, J.M.V. (1963), *Nature (London),* **200,** 1332–1333.
Robinson, P.J., Graf, L. and Sege, K. (1981), *Proc. Natl. Acad. Sci. U.S.A.,* **78,** 1167–1170.
Scheid, M. and Triglia, D. (1979), *Immunogenetics,* **9,** 423–433.
Siadak, A.W. and Nowinski, R.C. (1980), *J. Immunol.* **125,** 1400–1401.
Smithies, O. and Poulik, M.D. (1972), *Science,* **175,** 187–189.
Snell, G.D. (1948), *J. Genet.,* **49,** 87–101.
Snell, G.D. (1958), *J. Natl. Cancer Inst.,* **21,** 843–877.
Snell, G.D. and Bunker, H.P. (1965), *Transplantation,* **3,** 235–252.
Snell, G.D., Cherry, M., McKenzie, I.F.C. and Bailey, D.W. (1973), *Proc. Natl. Acad. Sci. U.S.A.,* **70,** 1108–1111.
Stanton, T.H., Bennett, J.C. and Wolcott, M.J. (1975), *J. Immunol.,* **115,** 1013–1017.
Steck, T.L. (1974), *J. Cell Biol.,* **62,** 1–19.
Taylor, B.A. (1980), in *Genetic Variants and Strains of the Laboratory Mouse* (M.C. Green and G. Fischer, eds), G.F. Verlag, Stuttgart.

Trowbridge, I.S. (1978), *J. Exp. Med.*, **148,** 313–323.
Trowbridge, I.S. and Mazauskas, C. (1976), *Eur. J. Immunol.*, **6,** 557–562.
Trowbridge, I.S., Ralph, P. and Bevan, M.J. (1975), *Proc. Natl. Acad. Sci. U.S.A.*, **72,** 157–161.
Vitetta, E.S., Uhr, J.W. and Boyse, E.A. (1975), *J. Immunol.*, **114,** 252–254.
Vitetta, E.S., Poulik, M.D., Klein, J. and Uhr, J.W. (1976), *J. Exp. Med.* **144,** 179–192.

2 Genetics of the Human Red Cell Surface

PATRICIA TIPPETT

EDITOR'S INTRODUCTION

For the whole of the 20th century, the red blood cell surface has been a favourite tool of geneticists and biochemists. One individual can be immunized to produce antibodies against the red cells of another by cross-reacting environmental antigens (this is the case with 'naturally occurring' antibodies to A and B blood groups), pregnancy (e.g. Rhesus antibodies) and accidentally during blood transfusions. All these sources provide a rich supply of antisera. The agglutination of red blood cells by antibodies, either directly or indirectly with the help of a second antibody against human immunoglobulin, provides a quick, cheap and simple assay. Combined with the ease of phlebotomy (the removal of blood) it can be seen why the human red blood cell surface has been more exhaustively studied than any other and, in consequence, the serological, genetic and biochemical definition of the red blood cell surface represents a model system for analysing other cell surfaces. In addition the medical importance of blood transfusion and of diseases such as haemolytic disease of the newborn ensures that study of the human red blood cell surface is not only an esoteric pursuit.

In Chapter 2, Patricia Tippett describes the enormous wealth of knowledge of the human red blood cell groups and explains how this information has been found to be useful for studying such diverse problems as malaria and chromosomal non-disjunction during meiosis.

Abbreviations

GalNac	*N*-acetyl-D-galactosamine
Gal	D-galactose
Fuc	L-fucose
GlcNac	*N*-acetyl-D-glucosamine
SDS/PAGE	sodium dodecyl sulphate/polyacrylamide gel electrophoresis
SGP	sialoglycoprotein
HDN	haemolytic disease of the newborn
HTR	haemolytic transfusion reaction
AIHA	autoimmune haemolytic anaemia

Genetic Analysis of the Cell Surface
(*Receptors and Recognition*, Series B, Volume 16)
Edited by P. Goodfellow
Published in 1984 by Chapman and Hall, 11 New Fetter Lane, London EC4P 4EE

2.1 INTRODUCTION

In 1901 Landsteiner observed that the red cells of some people were agglutinated by the sera of other people. This was the first definition of a blood group system and was eventually called ABO. Specific blood group antibodies now define more than 300 inherited red cell antigens; about three-quarters of these fall into 22 well-defined and genetically independent polymorphic blood group systems (those in Tables 2.1 and 2.9 and Chido:Rodgers). Blood groupers use Bodmer and Cavalli-Sforza's (1976) definition of polymorphism 'the occurrence of two or more alleles for a given locus in a population, where at least two alleles appear with frequencies of more than 1 per cent.' Most systems like ABO, MNSs and Rh (Rhesus) are polymorphic in all populations tested; other systems are polymorphic only in selected populations, for example Ge in Melanesians and LW in Finns. Other antigens are not polymorphic but classed, because of their incidence, as very high- or very low-frequency antigens. Yet other antigens are monomorphic, present on everyone's cells.

All red cell antigens to be described have been identified using manual and automated haemagglutination techniques. Polymers such as albumin, proteases which include papain, trypsin and ficin, and low-ionic strength solutions are often used to enhance direct agglutination. Indirect agglutination is achieved by using anti-human antibody reagents (antiglobulin test), the test devised by Coombs *et al.* (1945) for the detection of antibodies which coated cells but did not agglutinate them.

The first red cell antigens, A and B, were revealed by 'naturally occurring' antibodies; later systems (MN and P) were found as the result of deliberate immunization of animals with human red cells. All these antibodies agglutinated red cells. Since 1945 most 'new' red cell antigens have been found with immune antibodies reacting by the antiglobulin test. These alloantibodies are made by people immunized by an antigen, lacking from their genetic constitution, during transfusion or pregnancy. The antibody is not harmful to the mother immunized by transplacental passage of incompatible foetal cells but, if it is immunoglobulin G (IgG), the antibody can pass through the placenta and affect the foetus causing haemolytic disease of the newborn (HDN). Some of the same specificities are also found as autoantibodies, i.e. antibodies against antigens present on the patient's own red cells.

Other reagents used in classifying red cell antigens are lectins, extracts of seeds and snails, and, more recently, monoclonal antibodies. So far monoclonal antibodies have derived from mice immunized with human red cells, membrane extracts, blood group glycoproteins or lymphoid cell lines and provide only a few reagents distinguishing recognized antigens of the polymorphic systems.

The ABO antigens are widely distributed in many tissues; other antigens such as Rh and Xg^a are exclusive to red cells. Plasma antigens such as Lewis, Chido and Rodgers and some HLA antigens of leucocytes are also expressed on red cells.

The number of antigen sites per red cell varies from system to system: for ABO it is in the order of 250000 to 1×10^6, for Rh between 10000 and 40000 and for K less than 5000. The chemical nature of the antigens of ABO, Lewis and P are unequivocally established (Watkins, 1980) and some biochemistry of I-i and MNSs antigens is known.

The large number of recognized red cell antigen reflects the availability of red cells which are easier and less expensive to investigate than white cells, serum proteins or other tissues: a sample, enough for extensive tests, can be painlessly collected and does not deteriorate for 2–3 weeks. Clinically ABO and Rh are the most important systems. In ABO the naturally occurring antibodies can cause intravascular haemolysis if a patient is given ABO-incompatible blood. If Rh D-negative patients are transfused with D+ blood, 80% make anti-D. Recognition of the importance of ABO groups and the immunogenicity of D was the foundation of safe transfusion. Before recent prophylactic treatment of Rh-negative mothers the Rh antigen D was the most frequent cause of severe HDN.

Blood group antigens have no known biological function although there is association between some antigens and disease; Kell and chronic granulomatous disease, Duffy and malaria due to *Plasmodium vivax,* Wr^b and malaria due to *Plasmodium falciparum,* Rh_{null} and hereditary stomatocytic haemolytic anaemia.

Most red cell antigens are controlled by genes that behave as autosomal Mendelian dominants; this has been confirmed in thousands of families. However, blood group antigens are not usually primary gene products and their expression on the red cell surface may depend on more than one locus. Expression of red cell antigens is usually unaffected by age or environment and their precise inheritance makes them excellent tools for forensic, linkage and population studies. The frequency of antigens varies according to populations; some antigens are characteristic of certain races. In most blood group systems study of rare phenotypes has led to a better understanding of the genetic background. The established systems and the frequencies of their alleles in an English population are shown in Table 2.1.

Blood groups have shown that many genetic phenomena described in other species also apply to man. Amorphic or silent genes which do not code for any detectable antigen are known in all well-defined systems. The expression of some blood group antigens, such as Rh, can be modified by unlinked (and also by linked) suppressor genes which are usually recessive. However, one dominant suppressor, *In(Lu)* affects antigens of three genetically independent systems; Lutheran, P_1 and Au^a. Complex systems (MNSs, Rh, Lutheran and Kell) probably demonstrate the result of misaligned crossing-over leading to

Table 2.1 Established blood group systems with most important antigens and gene frequencies of an English population

System	Antigens	Gene frequencies	System	Antigens	Gene frequencies
ABO	$\mathrm{A_1}$	A^1 0.2090	Kell	K	K 0.0457
	$\mathrm{A_2}$	A^2 0.0696		k	k 0.9543
	B	B 0.0612			
		O 0.6602		$\mathrm{Kp^a}$	Kp^a 0.0245
				$\mathrm{Kp^b}$	Kp^b 0.9755
MNSs	M	MS 0.2371	Secretor		Se 0.5233
	N	Ms 0.3054			se 0.4767
	S	NS 0.0709			
	s	Ns 0.3866			
			Lewis	$\mathrm{Le^a}$	Le 0.8156
				$\mathrm{Le^b}$	le 0.1844
P	$\mathrm{P_1}$	P^1 0.5401			
		P^2 0.4599			
			Duffy*	$\mathrm{Fy^a}$	Fy^a 0.4250
				$\mathrm{Fy^b}$	Fy^b 0.5570
Rh	D	CDe 0.4205		$(\mathrm{Fy^b})$	Fy^x 0.0160
	C	cDE 0.1411			Fy 0.0020
	c	cDe 0.0257			
	E	CDE 0.0024	Kidd	$\mathrm{Jk^a}$	Jk^a 0.5142
	e	Cde 0.0098		$\mathrm{Jk^b}$	Jk^b 0.4858
		cdE 0.0119	Colton	$\mathrm{Co^a}$	Co^a 0.9559
		cde 0.3886		$\mathrm{Co^b}$	Co^b 0.0441
		CdE 0.0000			
Lutheran	$\mathrm{Lu^a}$	Lu^a 0.0390	Xg	$\mathrm{Xg^a}$	Xg^a 0.6590
	$\mathrm{Lu^b}$	Lu^b 0.9610			Xg 0.3410

* Canadian whites

hybrid genes or to gene duplication which may be followed by mutation. The blood group loci assigned to specific chromosomes are summarized in Table 2.2.

It is important to distinguish genotypes from phenotypes; this is achieved by italicizing *genes* and *genotypes* but not antigens and phenotypes.

Red cell system notation is inconsistent, having developed to accommodate different antigens over many years. In the ABO system, O is given a capital letter even though it is the result of an amorphic gene recessive to *A* and *B*. In later systems the second allele at a locus was given the small letter of the first allele, such as *k* or *s*, although in time a specific antibody, anti-k or anti-s, was found showing *k* and *s* to be co-dominant with *K* and *S* respectively. For other systems a suitable abbreviation of an appropriate name with a superscript 'a' was used for the first allele. Lu^a was the first antigen of the Lutheran system; phenotypes were written as Lu(a+) or Lu(a−) and the gene as *Lu^a*. The unidentified allele was called *Lu* until an antibody, antithetical to anti-Lu^a, was found promoting *Lu* to *Lu^b*.

Efforts to standardize notation by numbering are being organized by the International Society of Blood Transfusion. Although numbers are ideal for communication with computers, their assignment in complex systems is not straightforward and the genetic implications of a series of numbers is not immediately apparent.

Table 2.2 Chromosome assignments of blood group loci

Structural locus	Chromosome	SRO	Status
Fy	1	q12 → q21	C
Rh	1	p36 → p32	C
Sc	1	p34 → p32	C
Rd	1	p34 → p22.1	P
Jk	2		C
MNSs	4	q28 → q31	C
Ch, Rg	6	p23 → p2105	C
ABO	9	q34	C
Le	19		*
Se–Lu	19		*
Xg	X	p22.3 → pter	C
Background locus			
Xk	X		

From Human Gene Mapping 6 (1982): all through family studies. SRO = shortest region of overlap. p = short arm, q = long arm, ter = terminal (end of chromosome). C = confirmed assignment.
P = provisional assignment.
* Since Human Gene Mapping 6.

In this chapter the blood group systems are not presented chronologically but are divided into those with carbohydrate antigens, those associated with sialoglycoprotein, those which are complex etc. More details of the antigens and systems are available elsewhere (Race and Sanger, 1975; Issitt, 1979). Many references which are available in these and in a recent publication (Mollison, 1983) are not given here.

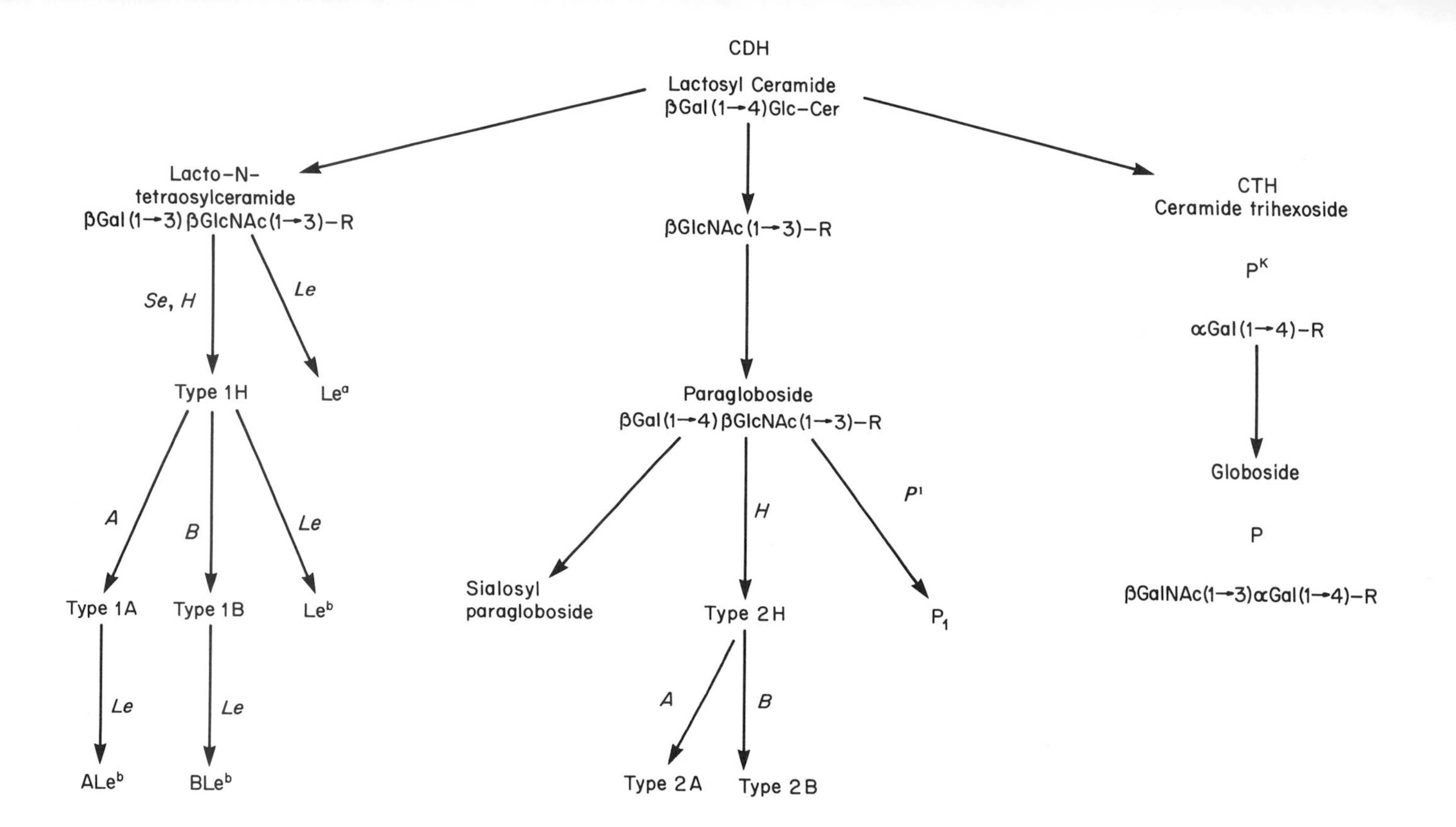

Fig. 2.1 Biosynthetic pathways showing relationship of ABH, Lewis, P_1, P^K and P glycolipids. For abbreviations and structures of the terminal immunodominant oligosaccharides, see Table 2.4.

2.2 SYSTEMS WITH CARBOHYDRATE ANTIGENS

The systems described in this section share the same biochemical background (Fig. 2.1), although they are genetically independent. Their antigens are carbohydrate and cannot therefore be primary gene products. The controlling genes code for glycosyltransferases that catalyse the transfer of a sugar to an acceptor substrate. Study of the rare O_h, Bombay, phenotype culminated in the understanding of the biosynthetic pathways of ABO and the appreciation that the co-operation of two loci (*ABO, Hh*) were involved in the expression of ABH on red cells and three loci (*ABO, Hh* and *Sese*) for their expression in secretions.

2.2.1 ABO system and Hh

The ABO groups depend on antigens present on red cells and the occurrence of reciprocal antibody in plasma. Any departure from this pattern in samples from adults heralds a rare phenotype, a chimera or, perhaps, a disease change. The six common patterns are shown in Table 2.3.

Table 2.3 Common ABO phenotypes and two rare phenotypes

Red cells						Serum					Saliva
Anti-						Cells					
A,	B,	A,B,	A_1	Phenotype	Genotype	A_1	A_2	B	O	Enzyme	if secretor
−	−	−	−	O	*OO*	+	+	+	−	H	H
+	−	+	+	A_1	A^1O, A^1A^1, A^1A^2	−	−	+	−	A, H	A, H
+	−	+	−	A_2	A^2O, A^2A^2	−*	−	+	−	A, H	A, H
−	+	+	−	B	*BO, BB*	+	+	−	−	B, H	B, H
+	+	+	+	A_1B	A^1B	−	−	−	−	A, B, H	A, B, H
+	+	+	−	A_2B	A^2B	−*	−	−	−	A, B, H	A, B, H
+	w	+	w	*cis* A_1B	A_1BO	−	−	w	−	A (B) H	A (B) H
−	−	−	−	O_h		+	+	+	+	**	

* Sometimes +.
**See text.

In addition to human reagents, lectins made from seeds (anti-A_1 *Dolichos bifloris* and anti-H *Ulex europaeus*) and from snails (*Helix hortensis* and *Helix pomatia*) are useful in ABO grouping. More recently, mice immunized with various human tissues or blood group substances have provided monoclonal anti-A (Voak *et al.*, 1980) or monoclonal anti-B (Sacks and Lennox, 1981) which are very useful as grouping reagents.

Samples from babies may lack ABO antibodies, any present are usually of maternal origin; the A_1 antigen is not always fully expressed at birth. A_2 cells have fewer A sites than A_1 cells; this probably reflects a true quantitative difference since the same A determinants are present on red cells and in secretions of A_1 and A_2 people.

Numerous variant alleles produce weak forms of A or weak forms of B (Race and Sanger, 1975; Salmon and Cartron, 1977; Watkins, 1980). One example producing weak B antigen and anti-B in plasma is characteristic of the rare *cis* AB phenotype (Table 2.3) in which A and B are inherited together from one parent. This exception to the usual mode of inheritance could be explained by intragenic crossing-over, by mutation in or near the active enzyme site or by gene duplication followed by mutation.

The rare O_h, Bombay, phenotype is ascertained because O cells are agglutinated by the anti-H present in the plasma (Table 2.3). Investigation of O_h propositi, and their families, led to the recognition of the *Hh* locus which controls production of H, the precursor of A and B. Lack of H, as in O_h people, prevents the expression of *A* and *B* (Fig. 2.1).

Hh segregates independently of *ABO*, showing that two independent loci are involved in the expression of the ABH antigens. A, B and H antigens are widely distributed as surface antigens on many tissues in the body and in 80% of the population are also secreted in a water-soluble form. The study of secreted substances, particularly from ovarian cysts, by many biochemists (schools of Morgan and Watkins, and of Kabat) led to the determination of the chemical constitution of these substances. These antigens are oligosaccharides with characteristic immunodominant sugars: *N*-acetyl-D-galactosamine for A and D–galactose for B. The oligosaccharides derive from two basic carbohydrate chains. Type 1 chain has a Gal(1→3)β-GlcNAc linkage and Type 2 chain has a Gal(1→4)β-GlcNAc linkage (Fig. 2.1) which can occur in highly branched chains. The simplest forms are shown in Table 2.4.

The ABH antigens are found as glycoproteins in secretions both as Type 1 and Type 2 determinants, and as glycosphingolipids in plasma as Type 1 determinants; on the red cell membrane the majority of ABH antigenic sites occur as glycoprotein and the rest as glycolipid. Red cells synthesize Type 2 oligosaccharides only; Type 1 determinants can be absorbed from plasma.

The H-specific enzyme, an α-2-L-fucosyltransferase, is absent from the serum of people of O_h phenotype but present in that of all other people. The B-specific enzyme, an α-3-D-galactosyltransferase, attaches Gal to H substance masking the H antigen. Addition of GalNAc to H substance to make A antigen is controlled by α-3-*N*-acetylgalactosaminyltransferase. The A-specific enzyme from A_1 people is distinguished from that of A_2 people by different pH optima (pH 6 for A_1, pH 7–8 for A_2) and isoelectric points (pI 9.9 for A_1, pI 6.9 for

Table 2.4 The terminal oligosaccharides of some red cell antigens

Antigen	Terminal oligosaccharides	Genes involved
Type 1 H	βGal(1→3)βGlcNAc-R ↑ 1,2 αFuc	*Se, H*
Type 1 A	αGalNAc(1→3)βGal(1→3)βGlcNAc-R ↑ 1,2 αFuc	*Se, H, A*
Type 1 B	αGal(1→3)βGal(1→3)βGlcNAc-R ↑ 1,2 αFuc	*Se, H, B*
Le^a	βGal(1→3)βGlcNAc-R ↑ 1,4 αFuc	*Le*
Le^b	βGal(1→3)βGlcNAc-R ↑ 1,2 ↑ 1,4 αFuc αFuc	*Se, H, Le*
A Le^b	αGalNAc(1→3)βGal(1→3)βGlcNAc-R ↑ 1,2 ↑ 1,4 αFuc αFuc	*Se, H, A, Le*
B Le^b	αGal(1→3)βGal(1→3)βGlcNAc-R ↑ 1,2 ↑ 1,4 αFuc αFuc	*Se, H, B, Le*
Type 2 H	βGal(1→4)GlcNAc-R ↑ 1,2 αFuc	*H*
Type 2 A	αGalNAc(1→3)βGal(1→4)βGlcNAc-R ↑ 1,2 αFuc	*H, A*
Type 2 B	αGal(1→3)βGal(1→4)βGlcNAc-R ↑ 1,2 αFuc	*H, B*
P_1	αGal(1→4)βGal(1→4)βGlcNAc-R	P^1
P^K	αGal(1→4)-R	
P	βGalNAc(1→3)αGal(1→4)-R	
R is Lactosylceramide	βGal(1→4)Glc-Cer	

A_2). Enzyme assays can distinguish A^1A^2 from A^1A^1 and A^1O. The O gene is a silent allele at the *ABO* locus not producing any transferase. O_h people, if genetically *A* or *B*, have A or B transferases in cells and in serum but no A or B antigens since they lack the precursor.

A and B transferases are found in plasma, red cell membranes, milk and many other tissues; H transferase is found in plasma and red cell membranes but its occurrence in other tissue depends on a person's secretor status. Investigation of twin chimeras showed that only about 20% of serum transferase derives from haemopoietic tissue.

The *ABO* locus is part of a linkage group with *Np* (Nail patella syndrome) and the enzyme AK_1 (adenylate kinase) and is on chromosome 9 (Table 2.2). Renwick and Schulze (1965) found that the *ABO* and *Np* linkage showed a sex difference in crossing-over rate (informative mothers had more cross-overs than informative fathers) at the same time as Cook (1965) reported a similar sex difference in the *Lu–Se* linkage.

2.2.2 ABH secretion

The occurrence of A, B and H substances in saliva is a dimorphic character dependent on genes at the *Secretor* locus which is independent of the *ABO* locus. The mechanism of the secretor genes is unknown: either a double dose of *se* suppresses formation of H transferase, or the *Se* gene may behave as a regulator gene allowing the expression of H transferase in secretory tissue.

The co-operation of *Se* with *Le* in forming the Lewis antigens expressed on red cells is discussed in the following section.

2.2.3 Lewis groups

The Lewis antigens, Le^a and Le^b, on red cells are not controlled by allelic genes but are the result of interaction of *H, Sese* and *Le* genes. They are not intrinsic to red cells but are secreted, as glycoprotein in saliva and as glycosphingolipid in plasma, and are secondarily adsorbed on to red cells. Le^a and Le^b antigens are carried on the same oligosaccharide chains as the ABH antigens.

There are two alleles at the Lewis locus: *Le*, an active gene, and *le*, a silent one. When present in the genome *Le* is always expressed; it is not under the control of the secretor gene. However, which antigen, Le^a or Le^b, is made depends on the genetic constitution of the *Hh* and *Sese* loci. As long as they are *HH* or *Hh,* people with at least one *Se* and one *Le* gene have Le(a−b+) red cells, are ABH secretors and Les (secretors of Le^a); people who are *se se* with at least one *Le* gene have Le(a+b−) red cells, are non-secretors of ABH but are Les; people who have at least one *Se* gene and are *le le* have Le(a−b−) red cells, are ABH secretors but are nL (do not secrete Le^a); *se se, le le* people have Le(a−b−) red cells, are non-secretors of ABH and nL.

Unlike most blood groups the Lewis groups are affected by age and by environment. The Lewis antigens are absent or very weak on the red cells of newborns, Le^a appears on the red cells of children with an *Le* gene and the adult phenotype is achieved by 6 years. Lewis antigens may be lost during pregnancy, reappearing after delivery. Transfused cells acquire the Lewis antigens of the recipient. Heterogeneity of Lewis antisera further complicates Lewis grouping, especially that of red cells.

The *Le* transferase uses a Type 1 chain since it adds L-fucose in a $1 \rightarrow 4$ linkage to a subterminal GlcNAc to make Le^a substance. Le^a antigen is not an acceptor for H transferase. If an *H* and a *Se* gene are also present L-fucose is added to the galactosyl as well as the GlcNAc residue to give Le^b substance (Table 2.4), thus *Le* may use A and B substances as acceptors to make complex antigens. *Le* transferase is found in milk and other tissues but not in plasma or serum.

Antibodies reacting with cells from *H, se se, le le* and from *H, Se se* or *Se Se, le le* people were inappropriately named anti-Le^c and anti-Le^d respectively, inappropriately because Le^c and Le^d are not coded for by alleles at the *Le* locus. Le^d is now recognized to be Type 1 H and Le^c is possibly an unsubstituted Type 1 chain (Table 2.4).

The Lewis system may be very important in renal transplantation (Oriol *et al.*, 1978; Williams *et al.*, 1978).

2.2.4 I-i blood groups

I antigen is found on red cells of the majority of adults but is absent from those of babies. The antigen is detected by cold allo and autoantibodies. Strong examples of anti-I from patients with cold agglutinin disease are usually monoclonal but have slightly different specificities from each other. I specificity is apparently dependent on branched-chain oligosaccharides, the various anti-I recognizing different inner portions of ABH active carbohydrate chains. So I antigens are precursors for the more complex ABH antigens (Feizi, 1981; Hakomori, 1981).

Sometimes in diseases, such as gastrointestinal adenocarcinoma, loss of transferases causes loss of A and B antigens and, therefore, increase in H, Le^b, I and i antigens.

I and i are not controlled by allelic genes, i behaves as a developmental antigen, strongly expressed at birth and diminishing until at 18 months old the adult level of I is achieved.

Families of rare adult i propositi, ascertained through their allo-anti-I, show that *I* is independent of *ABO, MNSs, P, Rh, Se, Fy* and *Jk*; it is not X- or Y-borne.

2.2.5 P system

The P system was found in 1927 by Landsteiner and Levine, was expanded in 1955 by Sanger's recognition of the association of P and the rare Tj(a−) phenotype, and now has five phenotypes defined by three antibodies (Table 2.5). P is now called P_1 and Tj(a−) is p.

Table 2.5 The P system defined by anti-P_1, −P and −P^K

Phenotype	Antigens on red cells	Antibodies in plasma	Approximate frequency in Europeans
P_1	P_1 P (P^K)	None	79%
P_2	P (P^K)	Anti-P_1*	21%
P_1^K	P_1 P^K	Anti-P	Very rare
P_2^K	P^K	Anti-P	Very rare
p	None	Anti-P_1PP^K	Very rare

* Sometimes present.
( ) Antigen too weak to be detected by direct agglutination.

P_1 is inherited as a Mendelian dominant; cells of P_2 people are negative with anti-P_1. Most people are P_1 or P_2, the rare P^K and p propositi always being ascertained through the antibody present in their plasma. The plasma of P^K donors, whether they be P_1^K or P_2^K, contains anti-P which reacts with all cells except those of P^K and p people (Table 2.5). The plasma of p people contains anti-P_1PP^K which reacts with all cells except those of p people; from some, but not all, anti-PP_1P^K sera anti-P^K can be isolated by adsorption with P_1 cells.

P_1 antigen is not fully developed at birth and its strength shows great variation in adults. The expression of P_1 is suppressed by the dominant inhibitor *In(Lu)*. The strength of P is very similar in P_1 and P_2 people, as is the strength of P^K in P_1^K and P_2^K people.

P^K behaves as a recessive antigen as judged by agglutination of red cells by anti-P^K. Family studies showed that the locus controlling the expression of P^K is independent of the P_1 locus.

The biochemistry of P_1, P and P^K is now well known. Morgan and Watkins first identified the P_1 antigen as a trisaccharide, αGal(1→4)βGal(1→4) GlcNAc, which they had isolated from the P_1P^K glycoprotein derived from sheep hydatid cyst fluid. They showed that anti-P^K was inhibited by a disaccharide Gal(1→4)Gal.

Naiki and Marcus (1974) using purified glycosphingolipids of known structure in inhibition tests made the surprising observation that P antigen is globoside and P^K is ceramide trihexoside. They isolated the P_1 glycolipid whose terminal trisaccharide was the same as that isolated from the glycoprotein.

They postulated the biosynthetic pathways (Fig. 2.1) for the P_1, P and P^K antigens whose structures are shown in Table 2.4.

Although the biochemistry is clear and undisputed, the genetic background is still undecided. Two pathways are involved: one producing P_1 and the other P through P^K. The biochemistry confirms that at least two and maybe three loci are involved. Isolation of the relevant α-galactosyltransferase should distinguish between possible genetic backgrounds.

Unfortunately the notation for the P system has proved unsuitable to accommodate the biochemical information; drastic revision is postponed until the genetic relationships between P_1, P^K and p are understood.

P antigens are found on cells other than red cells: P_1 and P on lymphocytes and fibroblasts, and P^K on fibroblasts. A P-like structure acts as a receptor for strains of *Escherichia coli* responsible for pyelonephritis. Källenius *et al.* (1981) suggest that p people may be resistant to infection by such bacteria. P antibodies are clinically important. Women of p phenotype are often ascertained by many early abortions; it is not yet known which antibody, anti-P_1, anti-P or anti-P^K, is responsible.

2.3 ANTIGENS ASSOCIATED WITH SIALOGLYCOPROTEIN

2.3.1 MNSs system

M and N were identified by Landsteiner and Levine in 1927 during their deliberate search for new human polymorphisms by immunizing animals. The system is now recognized to have two very closely linked loci, *MN* and *Ss*. The loci control the production of four common haplotypes and many variant and low-frequency antigens (Issitt, 1981). About 1% of Blacks are S−s− lacking both S and s antigens and, therefore, are U−.

The low-frequency antigens, identified by specific antibodies, are: Cl^a, Far(Kam), He, Hu, M_1, M^g, Mit, Mt^a, M^V, Ny^a, Ri^a, s^D, Sj, Sul, Vr and those of the Mi^a complex. Other variants are distinguished by their ability to react with some but not all antibodies of apparently the same specificity: M_2, N_2, S_z, S^B, M^c, M^r, M^z. Two antibodies anti-M^A and anti-N^A have subdivided M and N. Two antibodies react best when sialic acid is removed from the cell surface: anti-Can is anti-M-polypeptide related and anti-Tm is anti-N-polypeptide related. The chemistry of MNSs is bringing a little understanding of, and order to, this surfeit of antigens.

The M and N antigens are carried on the major sialoglycoprotein, α-SGP, also called glycophorin A. This transmembrane polypeptide was sequenced fully by Tomita *et al.* (1978); it has 131 amino acids with 15 sialic-rich *O*- and one *N*-glycosidically linked oligosaccharides. Different amino acids at positions 1 and 5 distinguish M (serine and glycine) from N (leucine and glutamic acid) (Table 2.6). Three of the *O*-glycosidically linked oligosaccharides occupy the peptide at amino acids 2, 3 and 4 and are necessary for the serologically determined antigens. Anstee (1981) suggested that the inherited antigenic differences reflect differences of conformation of α-SGP, effected by charge interactions between amino and carboxyl groups. This hypothesis explains the apparent destruction of M and N by neuraminidase treatment which removes sialic acid from the oligosaccharide. The terminal part of α-SGP is removed by trypsin so M, N and related antigens are lost.

Table 2.6 Amino acid sequence characteristic of some M and N antigens

	Amino acids 1 to 5
	α-SGP
N	Leu-Ser*-Thr*-Thr*-Glu-
M	Ser-Ser*-Thr*-Thr*-Gly-
M^g	Leu-Ser-Thr-Asn-Glu-
M^c	Ser-Ser*-Thr*-Thr*-Glu-
	δ-SGP
He	Trp-Ser*-Thr*-Ser*-Gly-Val-

* Glycosylated.

S and s are carried by δ, a minor SGP, also known as glycophorin B. Only part of this SGP has been sequenced: Dahr *et al.* (1980) found that at position 29 S has the amino acid methionine and s threonine. The amino acid sequence from positions 1 to 25 is identical with that of a blood group N α-SGP which explains the cross-reactivity of MM cells with anti-N. Cells which lack S and s, S−s−U−, lack normal δ. Dahr *et al.* (1980) suggest that carbohydrate is involved in the expression of S and s antigens. Nothing is known of the nature of the U antigen. The δ-SGP is not trypsin-sensitive but it is chymotrypsin-sensitive.

M^K, the silent gene, produces no M, N, S or s antigens and M^KM^K cells lack both α-SGP and δ-SGP.

Three of the rare variants at the MN locus have been sequenced (Table 2.6); two have substitutions in the α-SGP, the other affects the δ-SGP. M^g, which is identified by a specific antibody, has a single amino acid substitution at position 4 and is non-glycosylated. M^c which is not identified by a specific antibody has Ser, the M amino acid, in position 1 and Glu, the N amino acid in position 5. He, which is identified by a specific antibody, has substitutions at positions 1 and 5 in the Ss-SGP changing the normal 'N' activity to He.

Other rare antigens are associated with abnormal hybrid SGP and Anstee first applied the analogy between the Lepore and anti-Lepore haemoglobins to the MNSs SGP. These hybrid SGPs, which could have arisen from unequal crossing-over, and the antigens they carry are summarized in Table 2.7.

2.3.2 En^a

All En(a−) propositi were ascertained through antibody, usually immune, present in their plasma. At first En(a−) was thought to reflect homozygosity of a rare allele, *En*; now it is realized that En(a−) and related phenotypes can arise from several genetic backgrounds. The antibodies made by En(a−) people are heterogeneous, directed at different portions of the α-SGP.

Table 2.7 Hybrid sialoglycoproteins

SGP	Antigens on red cell		
Lepore type			
$(\alpha\text{–}\delta)^{MiV}$	M or N	s	Hil
$(\alpha\text{–}\delta)^{JR}$	(M)	S	
$(\alpha\text{–}\delta)^{AG}$	(M)	S	(Hil)
Anti-Lepore			
$(\delta\text{–}\alpha)^{Ph}$ with α	N	(s)	Dantu
$(\delta\text{–}\alpha)^{NE}$	N	(s)	Dantu
$(\delta\text{–}\alpha)^{St^a}$ with α and δ	M or N	S or s	St^a

(M) (S) or (Hil), antigen weaker than normal.

En En, En M^K, M^K M^K are all En(a−), lacking the major SGP α; related phenotypes lack part of the α-SGP because they only have a hybrid SGP of the 'Lepore' type (Table 2.7), for example homozygous *Mi^V* people. Lack, or reduction, of α-SGP explains all the reactions characteristic of En(a−) cells: depression of M and N antigens; lack of sialic acid which elevates the strength of the Rh antigens; increase in their agglutinability by animal sera and some lectins; reduction of electrophoretic mobility.

All En(a−) and related phenotypes are Wr(a−b−). *Wr^a* produces a low-frequency antigen and is genetically independent of *MN*. One donor, Mrs Fr., who was Wr(a+), made an antibody which was cautiously called anti-Wr^b. Ridgwell *et al.* (1983) using a monoclonal anti-Wr^b has suggested that Wr^b is on the trypsin-resistant portion on an α-helical region between residues 55 and 70 of α-SGP.

Miller *et al.* (1977) observed that En(a−) cells are resistant to invasion by *Plasmodium falciparum*. Pasvol *et al.* (1982) successfully pursued this observation, showing that α-SGP and δ-SGP are required for invasion but that the invasion may be blocked by coating normal red cells with monoclonal anti-Wr^b.

Tests on samples from 23 Tharu and 11 non-Tharu sent by Dr. Guido Modiano from Nepal, an area where *Plasmodium falciparum* is endemic, did not reveal any En(a−) samples.

2.3.3 Other antigens

T, Tn and Pr antigens are also expressed on SGP, but are not controlled by the *MNSs* locus.

(a) T and Tn

Most normal sera agglutinate cells which have been altered by certain enzymes produced by micro-organisms. This phenomenum is called polyagglutination and the antigens exposed are called cryptantigens. T, the first such antigen to be recognized, is expressed *in vivo* and *in vitro* if red cells are treated with neuraminidase. T was thought to be the precursor of M and N because M and N antigens were destroyed during the neuraminidase treatment which revealed T. It is now understood that the T antigen is the oligosaccharide chain which is exposed after removal of sialic acid (Anstee, 1981).

The immunodominant sugar of T antigen is β-galactose, removal of this sugar exposes Tn antigen. The different kinds of polyagglutination (T, Tn, etc.) can be distinguished by using selected lectins including *Arachis hypogaea, Dolichos biflorus* and *Salvia sclarea* extracts.

The apparent increase of T and Tn antigen in tissue from carcinoma of the breast and carcinoma of the colon (Springer *et al.*, 1979) may reflect deficiency of the sialo- or β-galactosyl-transferase required for addition of the next sugar, permitting the accumulation of the precursor substances. Reinforcement of the natural anti-T has been suggested as a possible method of immunotherapy.

(b) Pr

Pr is a monomorphic antigen, present on everyone's cells, which is inactivated by neuraminidase or protease treatment. It is detected by a rare cold agglutinin (Roelcke, 1974). All anti-Pr sera react with human red cells but are distinguished by their reactions with red cells from animals. Anti-Pr recognizes determinants on the oligosaccharides on both α- and δ-SGP (Anstee, 1981).

2.4 BLOOD GROUP SYSTEMS WITH COMPLEX LOCI

2.4.1 Rh

In 1939, Levine and Stetson reported an antibody, without naming it, in the serum of a mother whose baby had HDN. It was later thought to be the same as an antibody anti-Rh made by Landsteiner and Wiener in rabbits by injecting them with Rhesus monkey cells.

Two genetic backgrounds were suggested to accommodate the four related antigens known by 1943. Fisher's synthesis of Race's work postulated three closely linked genes, *DCE*, producing D or d, C or c and E or e which can be assembled in eight different gene complexes (haplotypes). These complexes are inherited in a straightforward way and their frequencies vary in different populations. The frequencies for an English population are shown in Table 2.1. Wiener supported an alternative theory that the Rh groups were produced by multiple alleles at a single locus (*Rh-Hr*). The notations for these two alternatives, together with the numerical notation of Rosenfield *et al.* (1962), are shown in Table 2.8. Rosenfield *et al.* (1962) devised their notation free from genetic implication, so that all recognized antigens could be individually reported. Over 40 Rh antigens are numbered (Table 2.8). Rosenfield *et al.* (1973), using numbers, present Rh as a conjugated operon with four units, each unit having an operator gene controlling the quantitative expression of its closely linked structural gene.

None of these theories completely fits all the facts and, although the genetic background is certainly more complex than three closely linked loci, CDE is the clearest language for communicating and interpreting the results of Rh

Table 2.8 Numerical Rh notation with CDE and Rh-Hr synonyms (after Race and Sanger, 1975)

CDE	Rh-Hr	CDE	Rh-Hr	CDE	Rh-Hr
Rh1 D	Rh_o	Rh15 *	Rh^C	Rh29 'Total Rh'	
Rh2 C	rh′	Rh16 *	Rh^D	Rh30 Go^a	
Rh3 E	rh″	Rh17 †	Hr_o	Rh31	hr^B
Rh4 c	hr′	Rh18	Hr	Rh32 $\bar{\bar{R}}^N$	
Rh5 e	hr″	Rh19	hr^s	Rh33 $R^{o\,Har}$	
Rh6 f, ce	hr	Rh20 VS, e^s		Rh34 Bas	Hr^B
Rh7 Ce	rh_i	Rh21 C^G		Rh35 1114	
Rh8 C^w	$rh^{w}{}_{1}$	Rh22 CE		Rh36 Be^a	
Rh9 C^x	rh^x	Rh23 Wiel, D^w		Rh37 Evans	
Rh10 V, ce^s	hr^V	Rh24 E^T		Rh38 Duclos	
Rh11 E^w	$rh^{w}{}_{2}$	Rh25 LW		Rh39 C-like	
Rh12 G	rh^G	Rh26 'Deal'		Rh40 Tar	
Rh13 *	Rh^A	Rh27 cE		Rh41 Ce-like	
Rh14 *	Rh^B	Rh28	hr^H	Rh42 Ce^s	

* Corresponding to some of the apparent anti-D made in D+people.
† Corresponding to an antibody made by -D-/-D-, etc.
Anti-Rh39 autoantibody, others alloantibodies.

testing and is sufficient for understanding the majority of Rh groups. After Fisher's synthesis more alleles at the three loci and antibodies to compound antigens were recognized but anti-d has not been found so *d* is considered an amorph.

There are many rare haplotypes which do not fit easily into the CDE concept. Some of these complexes are defined by specific antibodies; others are only detected, when appropriately partnered, by the ability of the red cells to react with some but not all antisera of apparently the same specificity. Only family studies, not serological tests, can distinguish between some of the inherited gene complexes and similar phenotypes caused by unlinked suppressor genes.

Family studies of the very rare Rh_{null} propositi who express no Rh antigens showed that two genetic backgrounds produce this unexpected phenotype. Two, or perhaps three, propositi are homozygous for a silent gene complex, – – –, producing no C, D or E antigens; these are described as the amorphic type of Rh_{null}. At least ten Rh_{null} propositi are of the 'regulator' type; their phenotype is caused by homozygosity of a rare suppressor gene which is independent of *Rh*. That Rh_{null} propositi of the regulator type, although not expressing Rh antigens, have *CDE* genes is often shown by the groups of their parents and children (Race and Sanger, 1975).

As well as lacking Rh antigens, both types of Rh_{null} have upsets of SsU groups and lack the high-frequency antigen, Fy5. Rh_{null} people often have a compensated haemolytic anaemia and their blood films show many stomatocytes. It is not surprising that Rh_{null} people have abnormal red cells since Rh is an integral part of the red cell membrane.

Rh_{null} people can, when immunized, make an antibody, anti-Rh29 which is called anti-'total Rh'. Monoclonal antibodies from mouse hybridomas of the same specificity have been identified in several laboratories. Using one such antibody Anstee and Edwards (1982) showed that 'total Rh' co-precipitated with band 3, or was carried on part of band 3.

Rh antigens have only been found on red cells not on other tissues. Despite many attempts to isolate Rh, the chemical nature of the antigens has, so far, proved elusive. Biochemists agree that Rh is lipoprotein but antigenicity was lost when attempts were made to isolate the protein. Hopefully the approach of Moore *et al.* (1982) of coating D+ cells with anti-D to stabilize the D antigen before SDS/PAGE treatment will yield some chemical information.

The D antigen is very immunogenic (80% of D– people transfused with D+ blood will make anti-D) so Rh grouping of patients requiring transfusion is essential to prevent Rh-negative recipients making anti-D. This is especially important for young girls and women of child-bearing age since anti-D, if IgG, will pass through the placenta and affect an Rh-positive foetus. Until recently some Rh-negative mothers were immunized by their Rh-positive babies to make anti-D which caused HDN, often severe, in later Rh-positive pregnancies. Rh immunoprophylaxis, injection of anti-D immunoglobulin to remove foetal D+ red cells from the mother's circulation, to suppress anti-D formation by D− mothers, and so save future children from HDN, is one of the success stories of the last decade (see Mollison, 1983).

Both Rh antigens and Rh antibodies are complex. Rh antigens are defined by human antibodies only, usually immune in origin, and such antibodies are multispecific. Hopefully human monoclonal antibodies, reacting with a single epitope, may lead to a better understanding of this complex group. One monoclonal anti-D, using Epstein-Barr virus transformation, has been produced and successfully cloned (Crawford *et al.*, 1983).

(a) LW blood group system

By 1963 it was proved that the animal anti-Rh made by Landsteiner and Wiener was different from the human anti-Rh (anti-D) which causes HDN. The Rh system as defined by human antibodies was widely established in an extensive literature, so the antigen defined by the animal antibody was renamed LW in honour of Landsteiner and Wiener.

Recently Sistonen and Tippett (1982) identified a second allele at the LW locus, LW^b, and showed *LW* to be polymorphic in Finland (Table 2.9). Both LW^a and LW^b are more strongly expressed in D+ than in D− cells of adults; cord cells regardless of D type give strong positive reactions. Although Rh and LW are phenotypically associated, *LW* is genetically independent of *Rh*. Both types of Rh_{null}, amorph and regulator, lack LW antigen.

2.4.2 Lutheran

This system is mainly defined by two co-dominant alleles, Lu^a and Lu^b (Table 2.1), a second sublocus codes for a low-frequency antigen Lu9 and its 'allelic' antigen Lu6. The null type, Lu(a−b−), has two genetic backgrounds distinguishable by family studies. One is dominant caused by an inhibitor gene, *In(Lu)*, and the other, a recessive character due to a silent gene, *Lu*, at the *Lutheran* locus.

Anti-Lu^a can be an immune or 'naturally' occurring antibody but is only rarely of clinical significance; anti-Lu^b is rare but has caused mild HDN. Lu(a−b−) propositi of the recessive type, *LuLu*, have all been ascertained through their immune antibody, anti-Lu3.

The numerical notation includes many antigens, which are detected by alloantibodies which fail to react with Lu(a−b−) cells. Race and Sanger coined the name para-Lutheran to describe those antigens which are phenotypically related to Lutheran but have not been shown to be coded by the *Lutheran* locus.

In(Lu), the dominant inhibitor of the expression of Lu^a and Lu^b, also inhibits the expression of P^1 and Au^a, genes independent of *Lu*. Lu(a−b−)

Table 2.9 Less-studied polymorphic blood group systems and their frequencies in Europeans

System	Gene frequencies	Not yet independent of‡	Remarks
Diego	*Di*^{a} 0.000 *Di*^{b} 1.000	Lu	Found in mongoloid races 36% Di(a+) in Carib Indians. Antibodies are clinically significant.
Cartwright	*Yt*^{a} 0.957 *Yt*^{b} 0.043	Di	Anti-Yt^{b} is rare.
Auberger	*Au*^{a} 0.692 *Au* 0.308	Se, Di, Yt	Suppressed by *In(Lu)*. Anti-Au^{a} is rare.
LW*	*LW*^{a} 0.971 *LW*^{b} 0.029	Lu, Le, Di, Yt, Au	LW(a−b−) can be inherited or acquired. Phenotypically related to D antigen of Rh (see Section 2.4.1a).
Dombrock	*Do*^{a} 0.420 *Do*^{b} 0.580	Au, LW	Usable anti-Do^{a} and anti-Do^{b} rare.
Sid	*Sd*^{a} 0.803 *Sd* 0.197	Lu, Le, Di, Yt, LW, Co	Antigen on red cells of adults very variable in strength, can be lost during pregnancy, absent from cord cells. Sd^{a} on other tissues and in body fluids. Related to Tamm-Horsfall glycoprotein.
Scianna†	*Sc1* 0.992 *Sc2* 0.008	Le, Di, Au, LW, Co, Sd	Sc:−1, −2 makes anti-Sc3. *Sc* on chromosome 1 (see Table 2.2).
Gerbich	all Ge+	Lu, Le, Di, Yt, Au, LW, Co, Sd, Sc	Polymorphic in Melanesians; complex system. In some families suppression of Kell antigens in Ge− members.
Indian	*In*^{a} 0.000 *In*^{b} 1.000	P, Rh, Lu, Le, Se, Fy, Di, Yt, Au, LW, Do, Co, Sd, Sc, Ge.	About 2–4% In(a+) in Indians.

* Polymorphic only in Finns.
† Frequency in Canadian donors.
‡ A 'new' system must be shown to be independent of those already known.

people of the dominant type also have less i antigen than do adults of common Lutheran phenotype. *In(Lu)* also inhibits the development antigens recognized by two monoclonal antibodies (Knowles *et al.*, 1982).

2.4.3 Kell

Kell was the first of many blood group antigens identified by the antiglobulin test when the cells of a baby were found to give a positive direct antiglobulin reaction which could not be attributed to Rh. Since 1946 the system has expanded; it is controlled by a complex locus composed of five subloci producing K or k, Kp^a or Kp^b or Kp^c, Js^a or Js^b, Wk^a(K17) or K11, and $U1^a$. The frequencies for *K*, *k*, Kp^a and Kp^b are shown in Table 2.1. The other uncommon alleles, Kp^c, Js^a, Wk^a and $U1^a$ are particular to certain races: Kp^c is found in 2% of Japanese, Js^a in 20% of Blacks, Wk^a in 0.3% of Northern Europeans and $U1^a$ in 3% of Finns. No haplotype with two uncommon alleles has been found.

People with the null phenotype, K_0, do not express any Kell antigens and, when immunized, make anti-Ku. Homozygosity for a suppressor or operator gene at the *Kell* locus which switches off all *Kell* structural loci is invoked to explain this rare phenotype.

Many alloantibodies to high-frequency antigens fail to react with K_0 cells. These para-Kell antigens (like the Lutheran ones) are phenotypically related to Kell but they have not yet been shown to be controlled by the *Kell* locus. Twenty-two Kell and Kell-related antigens have K numbers in the numerical notation originally proposed by Marsh.

Studies of another rare phenotype, McLeod, characterized by weakened Kell antigens, showed that a second locus, *Xk*, is involved in expression of Kell antigens. *Xk* which is X-borne (see Section 2.7.2) codes for Kx which behaves as the precursor for the *Kell* genes. K_0 cells which express no Kell antigens have elevated Kx.

Kell antigens are fully expressed at birth. K is very immunogenic. Anti-K can cause severe HDN and before cross-matching was common practice caused severe haemolytic transfusion reactions.

2.5 OTHER POLYMORPHIC BLOOD GROUP SYSTEMS

2.5.1 Duffy

Two alleles, Fy^a and Fy^b, originally defined this system. Sanger, Race and Jack found the majority of New York Blacks to be Fy(a−b−) which they attributed to a rare allele *Fy*. *Fy* has a frequency between 0.7 and 1.0 in different black populations and occurs rarely; the frequency is 0.002 in Whites. A fourth allele,

Fy^x, so far only observed in Whites, has a frequency of 0.016. *Fy*^{x} produces weak Fy^b antigen.

Fy may not be a silent allele, one antibody, numbered anti-Fy4 reacted with all Fy(a−b−) samples but insufficient serum was available for family studies.

Fy(a−b−) Blacks rarely make Duffy antibodies. An antibody which reacted with all cells that have Fy^a and/or Fy^b disclosed the first Fy(a−b−) white person. The antibody was numbered anti-Fy3.

Another antigen, Fy5, suggests an association between the genetically independent Rh and Fy systems; Rh_{null} cells as well as Fy(a−b−) cells are compatible with anti-Fy5.

Duffy is the only blood group which might show evidence of natural selection. Miller *et al.* (1975) showed that Fy(a−b−):−3 red cells, unlike those of other phenotypes, are not invaded *in vitro* by merozoites of *Plasmodium knowlesi* or *in vivo* by *Plasmodium vivax,* thus presumably accounting for the very high frequency of Fy(a−b−) in areas in which malaria was once endemic. The exact role of Fy^a/Fy^b/Fy3 antigens in malaria is not yet understood; merozoites attach to Fy(a−b−) cells but do not apparently invade them.

Duffy, the first blood group to be assigned to an autosome, joined chromosome 1 through linkage to the uncoiler locus, *1qh*. The most likely position for *Duffy* is 1q12→q21 (Table 2.2).

2.5.2 Kidd

Two alleles *Jk*^{a} and *Jk*^{b} (Table 2.1) define this system. A third allele, a silent one, *Jk* is very rare except in Polynesians. Homozygous *Jk Jk,* Jk(a−b−), when immunized, make an antibody, inseparable anti-Jk^{a}Jk^b (anti-Jk3). All Kidd antibodies anti-Jk^a, anti-Jk^b and anti-Jk3, are clinically significant and have been blamed for acute and delayed transfusion reactions as well as HDN.

Nothing is known of the chemical nature of the Kidd antigens. Perhaps the observation by Heaton and McLoughlin (1982) that Jk(a−b−) cells are resistant to lysis by 2M-urea may offer a clue to biochemists.

2.5.3 Colton

Two co-dominant alleles, *Co*^{a} and *Co*^{b}, define this system (Table 2.1). A third rare silent allele, *Co,* is postulated to explain the rare phenotype Co(a−b−). All the *Co Co* patients were found through their immune antibody, anti-Co^{a}Co^b (anti-Co3). Anti-Co^{a}Co^b is inseparable, it reacts with all cells which have Co^a and/or Co^b.

Not all examples of Co(a−b−) are due to homozygosity of the silent allele *Co*. The astonishing observation by de la Chapelle *et al.* (1975) that some patients with monosomy 7 in the bone marrow have the rare phenotype Co(a−b−): −3 is as yet unexplained.

2.5.4 Chido: Rodgers

Chido (Ch) and Rodgers (Rg) are primarily plasma antigens, determinants of complement C4d, controlled by two closely linked loci. Ch is C4B and Rg is C4A, as determined by starch-gel electrophoresis of neuraminidase-treated serum (Awdch and Alper, 1980). The frequency of both Ch−, C4BQO, and Rg−, C4AQO, is about 3%.

Anti-Ch and anti-Rg are immune antibodies made in response to transfusions of whole blood, plasma or plasma components (Hussain *et al.*, 1983). Anti-Ch does not cause transfusion reactions; Nordhagen and Aas (1979) reported a patient with anti-Ch who received nine pints of Ch+ blood uneventfully.

Ch and Rg are very weakly expressed on cells of newborn infants so anti-Ch and anti-Rg are not expected to cause HDN. The rare phenotype Ch−Rg− is only found in adults with C4 deficiency.

2.5.5 Less-studied blood group systems

Some other systems are less well defined either because reliable antisera identifying the antigens are not easily available or because the system is only polymorphic in some populations. These systems are briefly summarized in Table 2.9.

2.6 NON-POLYMORPHIC BLOOD GROUPS

Many red cell antigens are controlled by genes which are not polymorphic in any population yet tested. These are either very high-frequency (public) antigens controlled by genes with a frequency greater than 99% or very low-frequency (private) antigens with a frequency less than 1 in 400. Antibody in the serum of a patient lacking a public antigen presents an acute transfusion problem since such antibodies can cause haemolytic transfusion reactions (HTR) or haemolytic disease of the newborn (HDN). Antibodies to private antigens do not present transfusion problems but are often involved in HDN.

2.6.1 Public or high-frequency antigens

Phenotypes lacking the public antigens, shown to be inherited by having compatible sibs or family members, are summarized below.

Vel− The mode of inheritance is usually recessive but may in some families be dominant. Vel negatives have a frequency of about 1 in 2000 to 4000. The strength of Vel antigen in adults varies and that in cord samples is weaker than in adults. Anti-Vel has caused HTR.

Lan− Mode of inheritance is recessive. Lan negatives have a frequency of about 1 in 4000. Lan is well developed in cord bloods. Anti-Lan has caused HTR and HDN. Gn^a and So are the same as Lan.

Gy(a−) Mode of inheritance is recessive. Gy^a is weaker on cord cells than on those from adults. Anti-Gy^a can cause HTR.

Hy− Frequency of Hy− is about 1 in 10000. Related to Gy^a because all Gy(a−) are Hy− and white; other Hy− are weak Gy(a+) and black. Anti-Hy has caused HTR and HDN.

At(a−) All At(a−) are black. At^a is well developed at birth. Anti-At^a can cause HTR and HDN. El is the same as At^a.

Jr(a−) Mode of inheritance is recessive. Jr(a−) has a frequency of 1 in 4000 Japanese; it is rarer elsewhere. Jr^a is well developed at birth. Anti-Jr^a has caused HDN.

Jo(a−) Three unrelated Jo(a−) are all black.

JMH− Only one example is inherited, most examples are acquired (in elderly patients) or transient. Two other public antigens are also absent from JMH− cells. Anti-JMH is not clinically significant. A monoclonal antibody fails to react with JMH− (Daniels and Knowles, 1982).

Er(a−) Five unrelated examples.

Ok(a−) Only one example. Anti-Ok^a may be clinically significant.

Cr(a−) Several examples, usually black.

Tc(a−) Three examples, all black. Rare Cr(a−) Tc(a−) suggests relationship with Cr^a. Two low-frequency antigens Tc^b and Tc^c reported.

2.6.2 Private or low-frequency antigens

The private antigens controlled by the main blood group loci have appeared in appropriate sections; MNSs and Rh are both well endowed. All the antigens mentioned below are published, there are also many examples not yet published. Private antigens, when present, are excellent genetic markers.

Independence from all major blood group loci has been shown for Fr^a, Rd, Sw^a, Wr^a and from most loci for: An^a, Bp^a, By, Hov, Je^a, Jn^a, Ls^a, Mo^a, Pt^a, Rb^a, Rl^a, To^a, Wb, Wd^a, Wu. Less is known about: Bi, Bx^a, Chr^a, Dh^a, Good, Heibel, Hey, Li^a, Os^a, Pe, Poll, Re^a, Ts, Vg^a, Zd, Zt^a. Genes producing all these antigens are inherited as Mendelian dominants. Li^a may be linked to or even controlled by *Lu*.

Although antibodies to private antigens may cause HDN (e.g. anti-Bi, −By, −Good, −Heibel, −Li^a, −Poll, −Rd, −Wr^a, −Zd) most are naturally occurring often found in patients with AIHA or in people who have been hyperimmunized. Sera containing 'naturally occurring' antibodies usually contain antibodies to several antigens.

(a) Bg antigens

Some HLA antigens which are sometimes expressed on red cells, behaving like private antigens, are called Bg and cause problems during compatibility testing. The inheritance of Bg is not always straightforward since the same HLA antigen is not detected on red cells of other family members. HLA−B7 (Bg^a), −B17 (Bg^b), −A28 (Bg^c), −B8, −A9, −A10 and −B12 have been detected on red cells.

2.7 X-LINKED BLOOD GROUPS

The different phenotype frequencies in males and females and the characteristic pattern of inheritance in families makes easy the identification of dominant X-linked markers. Only two blood group loci are X-borne: *Xg*, a polymorphic system, and *Xk*, which is involved in the Kell system, perhaps controlling the precursor substance for Kell antigens.

2.7.1 The Xg blood groups

Anti-Xg^a, reported in 1962, identified the first blood group antigen controlled by a gene carried by a particular chromosome, the X. Xg^a is found only on red cells. Nothing is known of the chemical nature of Xg^a except that it is easily destroyed by proteolytic enzymes. Anti-Xg^a is not clinically significant.

The gene frequency of Xg^a, like that of other blood group genes, varies in different populations. So far New Guinea has the highest gene frequency for Xg^a (0.85) and Taiwan aborigines with 0.38 have the lowest. The gene frequencies of $Xg^a = 0.66$ and $Xg = 0.34$ in Northern Europeans give phenotype frequencies of 66% Xg(a+) in men and 88% in women. The identification of a benign X-linked gene of such useful frequency, about half Xg(a+) women are heterozygous Xg^aXg, was welcomed by geneticists interested in mapping the X-chromosomes.

The results of testing many thousands of samples from families with X-linked conditions disclosed very few loci within measurable distance of *Xg*. That *Xg* failed to help map the relative positions of many X-markers was disappointing but is easily understood now that *Xg* is known to be located at the end of the short arm of X, Xp22.3 → Xpter. Race and Sanger (1975) summarized the lod scores ('lod' denotes 'log odds' i.e. 'log probability ratio') for X-linked characters and Xg. Only three loci were close enough and common enough to show direct measurable linkage (sum of lod scores >2 at some value of θ): ichthyosis (STS), ocular albinism and retinoschisis. Since 1975 one more locus, *Xk*, has joined the Xg linkage group.

Xg^a studies of people with X-aneuploidy and their families were more rewarding than the linkage work. In about a third of XO women (Turners syndrome) Xg^a can show if the single X is of maternal or paternal origin: Xg provided the first proof that the single X of Turners could be paternal in origin. When there is an extra X as in XXY Klinefelters syndrome, Xg^a can show, in some cases, at which stage of cell division the non-disjunction providing the extra X occurred. Accidents can happen at gametogenesis or post-zygotically. Tests on families of Klinefelters showed that XXY could arise from non-disjunction of the first meiotic division of spermatogenesis and that non-disjunction at both meiotic divisions of spermatogenesis can occur to make an XXYY zygote.

Xg was the first X-borne locus thought to escape inactivation. This claim was somewhat controversial until, 20 years later, several laboratories showed that the neighbouring *STS* locus also escaped inactivation.

The relationship of Xg^a to 12E7 antigen (Goodfellow and Tippett, 1981), which initiated the search for an active Y-borne gene, is summarized in the next section.

(a) 12E7

A monoclonal antibody, 12E7, raised to a T-cell line (Levy *et al.*, 1979) was found to recognize an antigen controlled by an X-borne gene (Goodfellow *et al.*, 1980). This antigen was also expressed on red cells where it showed variation of expression. All Xg(a+) people are high expressors of 12E7 antigen. Although Xg(a−) males can be high or low expressors, Xg(a−) females are always low expressors (Goodfellow and Tippett, 1981). These observations could be explained either by sex-limitation of 12E7 expression in the absence of Xg^a or by a locus *Yg* with high-level (Yg^a) and low-level (*Yg*) alleles on the Y chromosome homologous to *Xg* on the X, both Xg^a and Yg^a producing high expression of 12E7 antigen. Goodfellow and colleagues showed that the expression of 12E7 antigen is controlled independently by genes on the X, at Xp22.3, and on the Y (Goodfellow *et al.*, 1983; Goodfellow, 1983).

2.7.2 Xk

The observation by Giblett *et al.* (1971) that boys with chronic granulomatous disease (CGD) often had weakened Kell antigens started investigations of rare Kell phenotypes. Marsh *et al.* (1975) postulated *Xk*, an X-borne gene, controlling the production of Kx which can be thought of as the precursor for the antigens coded for by the *Kell* structural locus (Section 2.4.3). Kx is on leucocytes as well as red cells; Marsh *et al.* (1975) suggested four alleles to explain the observed phenotypes (Table 2.10).

Using routine Kell antisera the McLeod phenotype is: K-kw; Kp(a−bwc−); Js(a−bw); K:−11, −17; U1(a−); Ku w; Km−; Kx−. No-one with McLeod phenotype has yet been reported positive for a less-common allele at one of the subloci.

Table 2.10

	Kx on:			
	Red cells	Leucocytes	Kell groups	CGD
X^1k	Yes	Yes	Normal	No
X^2k	No	No	McLeod	Yes
X^3k	Yes	No	Normal	Yes
X^4k	No	Yes	McLeod	No

X^1k is the common gene, the other three (X^2k, X^3k, X^4k) are all rare. X^3k is more frequent that X^2k. X^2k: some boys with CGD lack Kx on red cells and leucocytes and when transfused make antibodies, anti-(Km+Kx). X^3k: boys with CGD but having normal Kell groups do not make anti-Kx when transfused. X^4k: males with McLeod phenotype who have normal Kx on their leucocytes are apparently normal but closer investigation has shown they have an acanthocytic haemolytic anaemia. People with McLeod phenotype without CGD have elevated CPK (creatine phosphokinase) with associated muscular myopathy (Marsh *et al.*, 1981), and recently Schwartz *et al.* (1982) have reported that these patients also have arreflexia.

The *Xk* locus is subject to Lyonization. Red cells of some mothers of boys with McLeod phenotype often give a mixed field pattern of agglutination with Kell antisera showing that only some cells express Kell antigens. If *Xk* is inactivated no Kx is available for the *Kell* structural genes and so a greatly reduced amount of Kell antigens is produced.

Xk is closely linked to *Xg*. Only one possible recombinant is reported; it is only possible because the propositus could be mutant or his McLeod phenotype caused by a different background.

2.8 CONCLUSIONS

The red cell membrane is well characterized by many inherited polymorphic and non-polymorphic determinants and also some monomorphic antigens. Studies of these antigens have contributed to the safety of blood transfusion and to problems of identity. The recent flowering of monoclonal antibodies promises recognition of many more markers, especially of the monomorphic type, and insight into the development and biochemistry of antigens already known.

Null phenotypes in the established systems illustrate different genetic backgrounds. The most common is homozygosity of a silent allele at the structural locus, as is seen in Rh, Kell, Lu, Fy and Jk. Other null phenotypes reflect homozygosity of a recessive allele suppressing antigens produced by an independent locus as seen in ABO, Rh and Kell; interference in the biosynthetic pathway of ABO is proved and for Kell is postulated. One dominant suppressor, *In(Lu)*, affects the expression of antigens produced by three genetically independent systems *Lu*, P^1 and *Au*.

REFERENCES

Anstee, D.J. (1981), *Semin. Hematol.*, **18,** 13–31.

Anstee, D.J. and Edwards, P.A.W. (1982), *Eur. J. Immunol.*, **12,** 228–232.

Awdeh, Z.L. and Alper, C.A. (1980), *Proc. Natl. Acad. Sci. U.S.A.*, **77,** 3576–3580.

Bodmer, W.F. and Cavalli-Sforza, L.L. (1976), *Genetics, Evolution, and Man,* Freeman, San Francisco, p.763.

Cook, P.J.L. (1965), *Ann. Hum. Genet.*, **28,** 393–401.

Coombs, R.R.A., Mourant, A.E. and Race, R.R. (1945), *Br. J. Exp. Pathol.*, **26,** 255.

Crawford, D.H., Barlow, M.J., Harrison, J.F., Winger, L. and Huehns, E.R. (1983), *Lancet,* **i,** 386–388.

Dahr, W., Beyreuther, K., Steinbach, H., Gielen, W. and Krüger, J. (1980), *Hoppe-Seyler's Z. Physiol. Chem.*, **361,** 895–906.

Daniels, G.L. and Knowles, R.W. (1982), *J. Immunogenet.*, **9,** 57–59.

de la Chapelle, A., Vuopio, P., Sanger, R. and Teesdale, P. (1975) *Lancet,* **ii,** 817.

Feizi, T. (1981), *Immunol. Commun.*, **10,** 127–156.

Giblett, E.R., Klebanoff, S.J., Pincus, S.H., Swanson, J., Park, B.H. and McCullough, J. (1971), *Lancet,* **i,** 1235–1236.

Goodfellow, P. (1983), *Differentiation* **23**(Suppl.), 35–39.

Goodfellow, P.N. and Tippett, P. (1981), *Nature (London),* **289,** 404–405.

Goodfellow, P., Banting, G., Levy, R., Povey, S. and McMichael, A. (1980), *Somat. Cell. Genet.*, **6,** 777–787.

Goodfellow, P., Banting, G., Sheer, D., Ropers, H.H., Caine, A., Ferguson-Smith, M.A., Povey, S. and Voss, R. (1983), *Nature (London),* **302,** 346–349.

Hakomori, S-I. (1981), *Semin. Hematol.*, **18,** 39–62.

Heaton, D.C. and McLoughlin, K. (1982), *Transfusion,* **22,** 70–71.

Human Gene Mapping 6 (1982), *Cytogenet. Cell Genet.*, 32.

Hussain, R., Edwards, J.H., Rizza, C.R. and Tippett, P. (1983), *Lancet,* **i,** 585.

Issitt, P.D. (1979), *Serology and Genetics of the Rhesus Blood Group System,* Montgomery Scientific Publications, Cincinnati, Ohio.

Issitt, P.D. (1981), *The MN Blood Group System,* Montgomery Scientific Publications, Cincinnati.

Källenius, G., Svenson, S.B., Möllby, R., Cedergren, B., Hultberg, H. and Winberg, J. (1981), *Lancet,* **ii,** 604–606.

Knowles, R.W., Bai, Y., Lomas, C., Green, C. and Tippett, P. (1982), *J. Immunogenet.*, **9,** 353–357.

Levy, R., Dilley, J., Fox, R.I. and Warnke, R. (1979), *Proc. Natl. Acad. Sci. U.S.A.*, **76,** 6552–6556.

Marsh, W.L., Øyen, R., Nichols, M.E. and Allen, F.H. (1975), *Br. J. Haematol.*, 247–262.

Marsh, W.L., Marsh, N.J., Moore, A., Symmans, W.A., Johnson, C.L. and Redman, C.M. (1981), *Vox Sang.*, **40**, 403–411.

Miller, L.H., Mason, S.J., Dvorak, J.A., McGuinniss, M.H. and Rothman, I.K. (1975), *Science*, **189**, 561–563.

Miller, L.H., Haynes, J.D., McAuliffe, F.M., Shiroishi, T., Durocher, J.H. and McGuinniss, M.H. (1977), *J. Exp. Med.*, **146**, 277–281.

Mollison, P.L. (1983), *Blood Transfusion in Clinical Medicine*, 7th edn. Blackwell Scientific Publications, Oxford.

Moore, S., Woodrow, C.F. and McClelland, B.L. (1982), *Nature (London)*, **295**, 529–531.

Naiki, M. and Marcus, D.M. (1974), *Biochem. Biophys. Res. Commun.*, **60**, 1105–1111.

Nordhagen, R. and Aas, M. (1979), *Vox Sang.*, **37**, 179–181.

Oriol, R., Cartron, J., Yvart, J., Bedrossian, J., Duboust, A., Bariety, J., Gluckman, J.C. and Gagnadoux, M.F. (1978), *Lancet*, **i**, 574–575.

Pasvol, G., Jungery, M., Weatherall, D.J., Parsons, S.F., Anstee, D.J. and Tanner, M.J.A. (1982), *Lancet*, **ii**, 947–951.

Race, R.R. and Sanger, R. (1975), *Blood Groups in Man*, 6th edn., Blackwell Scientific Publications, Oxford.

Renwick, J.H. and Schulze, J. (1965), *Ann. Hum. Genet.*, **28**, 379–392.

Ridgwell, K., Tanner, M.J.A. and Anstee, D.J. (1983), *Biochem. J.*, **209**, 273–276.

Roelcke, D. (1974)., *Clin. Immunol. Immunopathol.*, **2**, 266–280.

Rosenfield, R.E., Allen F.H., Swisher, S.N. and Kochwa, S. (1962), *Transfusion*, **2**, 287–312.

Rosenfield, R.E., Allen, F.H. and Rubinstein, P. (1973), *Proc. Natl. Acad. Sci. U.S.A.*, **70**, 1303–1307.

Sacks, S.H. and Lennox, E.S. (1981), *Vox Sang.*, **40**, 99–104.

Salmon, Ch. and Cartron, J.P. (1977), in *CRC Handbook Series in Clinical Science: Section D. Blood Banking Vol. 1* (T.J. Greenwalt and E.A. Steane, eds), CRC Press, Cleveland, pp. 69–256.

Schwartz, S.A., Marsh, W.L., Symmans, A., Johnson, C.L. and Mueller, K.A. (1982), *Transfusion*, **22**, 404.

Sistonen, P. and Tippett, P. (1982). *Vox Sang.*, **42**, 252–255.

Springer, G.F., Desai, P.R., Murthy, M.S. and Scanlon, E.F. (1979), *J. Surg. Oncol.*, **11**, 95–106.

Tomita, M., Furthmayr, H. and Marchesi, V.T. (1978), *Biochemistry*, **17**, 4756–4770.

Voak, D., Sacks, S., Alderson, T., Takei, F., Lennox, E., Jarvis, J., Milstein, C. and Darnborough, J. (1980), *Vox Sang.*, **39**, 134–140.

Watkins, W.M. (1980) *Adv. Hum. Genet.* **10**, 1–136.

Williams, G., Pegrum, G.D. and Evans, C.A. (1978), *Lancet*, **i**, 878.

3 Analysis of the Human Cell Surface by Somatic Cell Genetics

ALAN TUNNACLIFFE and
PETER GOODFELLOW

EDITOR'S INTRODUCTION

Genetic analysis using classical Mendelian techniques suffers from several drawbacks when applied to *Homo sapiens*. First, humans have a long generation period. Second, because of personal choice in mating, laudable ethical constraints on experimenters and small family sizes, matings are frequently less informative than would be the case for experimental animals. Third, the paucity of genetic markers present at polymorphic levels (i.e. greater than 1%) in human populations also limits Mendelian analysis (however, the recent introduction of restriction fragment polymorphisms is beginning to obviate this last limitation). The solution to these problems was the introduction of somatic cell genetic analysis using human–rodent hybrids.

Hybrid cells produced by artificially fusing rodent and human cells with sendai virus or polyethylene glycol retain the rodent chromosomes but spontaneously segregate human chromosomes. Correlation of the presence of human markers with the presence of human chromosomes or chromosomal fragments forms the basis of rapid gene mapping by somatic cell genetics. The technique has the added advantage that many more phenotypic differences can be detected between species (i.e. between humans and rodents) than within species, thereby, greatly expanding the number of markers which can be studied.

Surprisingly, analysis of cell surface molecules by somatic cell genetics has lagged behind that of other markers. This is particularly strange because traditionally human genetics has employed cell surface markers (see Chapter 2), and cell surface molecules defined by antibodies have several advantages. In Chapter 3, Alan Tunnacliffe and I describe the methods and technology of somatic cell genetics.

Genetic Analysis of the Cell Surface
(*Receptors and Recognition*, Series B, Volume 16)
Edited by P. Goodfellow
Published in 1984 by Chapman and Hall, 11 New Fetter Lane, London EC4P 4EE

3.1 INTRODUCTION

Genetic analysis can be used to define the components and functions of complex systems. A preliminary to this analysis is the identification of the genes responsible and their linkage relationship to each other. This process is known as gene mapping and results in the assignment of a gene to an individual chromosome and the ordering of genes on each chromosome. The cell surface is a complex organelle, responsible for communication between the cell and its environment, which is amenable to genetic analysis. Many components of the human cell surface have been identified immunologically: coupling the immunological approach with standard somatic cell genetic techniques can lead to the chromosomal assignment of genes which govern cell surface biochemistry. In a complementary manner, genetically characterized cell surface components can be used to manipulate the genome of somatic cell hybrids.

The past decade has resulted in an explosion in the size of the human gene map: by 1981 over 300 genes had been mapped to the autosomes (Sixth International Workshop on Human Gene Mapping, Oslo 1981) and journals continue to publish further additions at an increasing rate. The reason for this burgeoning is the development of somatic cell genetics. Cells when fused together form, at low frequency, hybrid cells, which combine the chromosomes from both parental cells in one nucleus. Hybrid cells produced between human and rodent cells randomly lose human chromosomes. The presence of a human gene product can be correlated with the presence of a specific human chromosome thereby assigning the gene to that chromosome (see Fig. 3.1). Gene assignment requires a gene assay, or an assay for a product of that gene: it is also necessary that the human gene or gene product be distinguishable from the corresponding rodent gene in hybrids. The specificity of an enzyme for its substrate, or a related compound, coupled with the ability to separate many human and rodent isoenzymes by gel electrophoresis (Harris and Hopkinson, 1976) have allowed the chromosomal mapping of numerous enzyme-encoding loci.

Assays for cell surface receptors have traditionally involved the binding of a (labelled) ligand to the receptor, and although this satisfies the first requirement for a gene assay (i.e. specificity), it often fails to satisfy the second requirement (i.e. *species*-specificity). Thus, human transferrin will bind to both human and mouse transferrin receptors (Goodfellow *et al.*, 1982a). It is understandable, therefore, why the mapping of receptor-encoding loci has lagged behind the mapping of genes for enzymes. A possible solution to this problem involves the use of suitable mutants as rodent parent of hybrids and this is how the gene coding for the epidermal growth factor receptor (*EGFR*) was first mapped (Shimizu *et al.*, 1980; Davies *et al.*, 1980). Hybrids were made between various mouse lines, negative for EGF-binding activity, and human cells. Then, binding of labelled EGF to hybrids was correlated with the presence of human chromosome 7. However, the limited availability of mutants, the problem of a possible re-expression of the rodent gene and the need to determine the nature of the genetic lesion restrict the value of this approach.

The treatment of the cell surface component as an antigen has proved to be a valuable approach to the problem. Originally with conventional antisera and more recently with monoclonal antibodies, it has been shown that species-specific recognition of surface determinants by antibodies satisfies the criteria for gene mapping. In this review, we shall describe this approach and outline the techniques involved and show how they are applicable to systems familiar to the cell biologist and also how new systems may be defined.

3.2 HYBRID PRODUCTION AND CHARACTERIZATION

3.2.1 Choice of hybrid parents

The production of hybrids intended for use in gene mapping requires a careful choice of parent cells and there are two particularly important aspects. First, unless a DNA probe is available, there is the problem of tissue-restricted expression. For example, if the molecule of interest is only found on, say, the surface of fibroblast cells, then it is advisable to choose those cells as human parent. Equally important is the choice of rodent parent, since the rodent cell must allow expression of the human gene. Extinction of human gene expression can result if human and rodent parents are incompatible and, in the above example, it would be wise to choose, say, an L cell or 3T3 line as a mouse parent. Having said this, expression of a human gene, silent in the human parent cell, can sometimes be seen in hybrids if the rodent cell can reprogram that gene. For example, several human liver-specific enzyme functions were exhibited in hybrids between human skin fibroblasts or neuroblastoma cells and a rat hepatoma line which allowed mapping of two enzyme loci (Kielty *et al.*, 1982a,b).

Secondly, since characterizing hybrids will involve chromosome analysis, it is preferable that a human cell with a normal karyotype is used as hybrid parent since the presence of markedly rearranged chromosomes can make a definitive gene assignment difficult, although well-defined chromosome translocations or deletions are used extensively for regional assignments. It should be noted that most immortal human cell lines have abnormal karyotypes and, where possible, fresh human tissue or primary cultures should be used in fusions.

3.2.2 Cell fusion and hybrid selection

Having made the choice of parent cells, hybrids are made by fusion with polyethylene glycol (PEG) (Pontecorvo, 1975) or, less frequently nowadays,

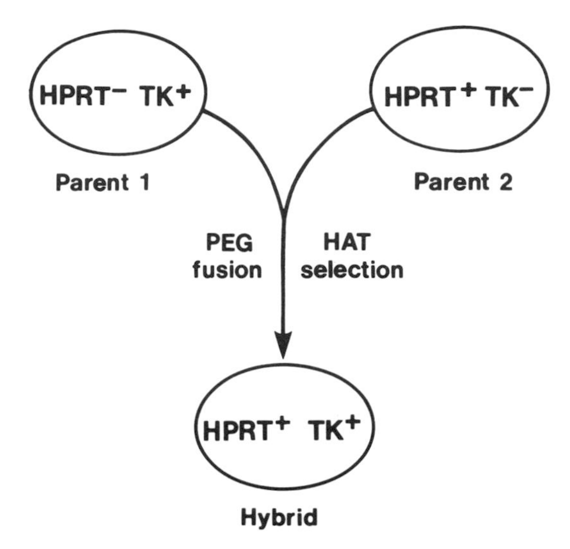

Fig. 3.1 Schematic production and selection of hybrids from $HPRT^-$ and TK^- parent cells by fusion with polyethylene glycol (PEG) and selection in HAT medium.

inactivated sendai virus (Okada, 1958, 1962; Harris and Watkins, 1965; Neff and Enders, 1968). Fusion conditions with PEG have been optimized since the technique was first introduced and have been reviewed extensively (Kennett *et al.*, 1981; Mercer and Baserga, 1982; Norwood and Zeigler, 1982). Hybrids are isolated by selection conditions which allow preferential growth of fusion products. The HAT selection system (Szybalski *et al.*, 1962) is most frequently employed for hybrid selection (Littlefield, 1964) and involves the use of mutant parents (Fig. 3.1): Parent 1 is $HPRT^-$ (hypoxanthine phosphoribosyltransferase; EC 2.4.2.8) and Parent 2 is TK^- (thymidine kinase, EC 2.7.1.21). Hybrids complement each others' defects and are $HPRT^+$ TK^+ allowing growth in HAT medium (hypoxanthine, aminopterin or methotrexate and thymidine), whereas parents are killed, since the activity of both enzymes is required: aminopterin inhibits dihydrofolate reductase and thus blocks *de novo* synthesis of purines and one-carbon-transfer reactions, including the conversion of deoxyuridylic acid to thymidylic acid. Cells growing in the presence of aminopterin cannot therefore survive, unless precursors for DNA synthesis are provided in the growth medium. In practice, when making hybrids for gene mapping, full selection is not always employed since 'normal' cells are often used as the human parent. (An exception to this is the use of fibroblasts from a patient with Lesch-Nyan syndrome, which are $HPRT^-$.) Thus, whilst mouse parents are either $HPRT^-$ or TK^- and are killed in HAT, selection against human parents is often due to growth properties: thymocytes and unstimulated lymphocytes will not grow in culture, whilst fibroblasts are usually outgrown by hybrids (Nabholz *et al.*, 1969). The increased resistance of rodent and hybrid cells to ouabain allows its use in selection medium to kill off human parent cells if desired (Kucherlapati *et al.*, 1975).

3.2.3 Determination of human chromosomes in hybrids

When continuous cultures of human–rodent somatic cell hybrids were first made, it was noted that human chromosomes were preferentially lost from hybrids, while the full rodent genome was retained (Weiss and Green, 1967). It is on this observation that the ability to map human genes in hybrids is based. Retention of human genetic material usually relies on selection for a single gene (*HPRT* or *TK*), and whilst this can ensure that either the human chromosomes carrying these genes (X or 17, respectively) are present in hybrids, other chromosomes are lost approximately randomly. Thus, different hybrids will have different numbers and distributions of human chromosomes and by performing gene assays on many hybrid clones, gene assignment is possible. This is illustrated in Section 3.4.2.

The initial instability of the karyotype of heterokaryons is followed by a period of relative stability and the human genetic complement of a hybrid can be constant for many cell generations, although ultimately we would expect that hybrids would lose all but the selected human chromosome or gene. Once a hybrid is established, we need to know which human chromosomes are present. The most reliable method involves karyotype analysis where spreads of mitotic hybrid cells are examined for human chromosomes. One standard procedure involves G11 staining (Bobrow and Cross, 1974) followed by quinacrine staining (Caspersson *et al.*, 1971) of the same spread. The G11 technique differentially stains human and rodent chromosomes, whilst quinacrine gives banding patterns characteristic of each human chromosome. Karyotype techniques are reviewed in Dev and Tantravahi (1982). These methods ensure that normal human chromosomes are present in hybrids and allow confidence in hybrids used for gene assignments. Hoever, karyotyping is labour- and time-intensive and requires highly skilled personnel. Other methods of determining the human contribution to hybrids involve the detection of markers specific for each of the human chromosomes. Since at least one enzyme has been mapped to each of the human chromosomes except Y, marker isozyme analysis is routinely used to detect human chromosomes in hybrids (reviewed in O'Brien *et al.*, 1982). This review highlights the use of antibodies in somatic cell genetics: one use is of a panel of monoclonal antibodies recognizing marker antigens for each of the human chromosomes. We are attempting to assemble such a panel (Section 3.4.2; Table 3.2) and so far have antibodies covering almost half of the human chromosomes. In the near future, it should be possible also to use cloned DNA fragments from each chromosome which hybridize to restriction fragments distinguishable from rodent counterparts. The drawback with the use of chromosome markers for hybrid analysis is that only one or a few genes (or restriction fragments) are assayed and if chromosome translocations or rearrangements have occurred in hybrids, these might not be detected. For this reason, karyotypic analysis is presently indispensable.

Having determined which human chromosomes are present in hybrids, gene assays can be carried out, but there are several caveats. (a) It is, of course, essential that hybrids are clones from separate culture vessels, although the appearance of subpopulations from a clonal isolate is almost inevitable. For this reason (b) periodic checks of the human chromosome content of hybrids should be carried out to detect any segregation which is occurring. This can conveniently be done by marker analysis. (c) Frozen stocks of hybrids should be made at regular intervals after characterization, or a large pool of cells grown up and characterized and then frozen in small aliquots. (d) Several independent sets of hybrids should be used for gene assignments to protect against possible genetic eccentricities of the human parent cell or undetected chromosomal rearrangement in hybrids.

3.3 ANTIBODY PRODUCTION AND CHARACTERIZATION

Antibodies as reagents for gene mapping need to be (a) specific for a particular gene product, and (b) species-specific or species-restricted, unless suitable rodent mutants are available. These criteria are achieved differently for conventional polyclonal antisera and monoclonal antibodies.

3.3.1 Polyclonal antisera

The production of useful polyclonal antibodies requires a substantially purified or genetically isolated immunogen. Animals are immunized according to simple protocols and subsequently bled. Antisera must then be adsorbed against rodent cells to remove any cross-reacting antibodies. This produces an antiserum which is species-specific for the gene product of interest. The technique of genetic isolation for the production of antisera against human cell surface determinants was developed by Buck and Bodmer (1974). A human–mouse hybrid with only a few human chromosomes is used to immunize a strain of mouse isogenic with that from which the mouse parent of the hybrid was derived. Thus, antibodies are raised against the human cell surface material only, unless the hybrid also expresses C-type virus tumour antigens. In this case, adsorption with hybrid parent mouse cells should remove contaminating antibodies. The value of the technique lies in the ability to produce antibodies against human material encoded by single chromosomes or chromosomal regions isolated from the remainder of the human genome. This is described in Section 3.4.1.

3.3.2 Monoclonal antibodies

Monoclonal antibody production (Kohler and Milstein, 1975; reviewed in Kennett *et al.*, 1981; Bastin *et al.*, 1982) in theory requires a less stringent purity of immunogen, since appropriate screening can isolate the desired antibody. Thus, whole human cells can be used to immunize mice and individual hybridomas screened. In practice, certain antibodies will predominate and it can be difficult to find a particular antibody since the antigen of interest may be present in only small amounts or lost in the immunological pecking order of a complex immunogen. Again, purification or genetic isolation of the antigen are helpful. We have used the latter method to produce monoclonal antibodies recognizing antigens encoded by the X (Hope *et al.*, 1982) and 11 (Tunnacliffe *et al.*, 1983; see Table 3.2) chromosomes. Abnormal expression or presentation of an antigen on the surface of a particular cell type may enhance the immunogenicity of the antigen. This phenomenon has been used to produce monoclonal antibodies against the EGF receptor (Waterfield *et al.*, 1982) and the insulin receptor (Roth *et al.*, 1982; Kull *et al.*, 1982) for example. Binding of monoclonal antibodies to human and rodent cells measured by radioimmunoassay or ELISA will determine whether the cognate antigenic determinant is species-specific. Strictly, this does not demonstrate true species-specificity, but merely a polymorphism between comparable molecules on human and rodent cell surfaces. It is often found that monoclonal antibodies directed against human determinants which do not react with mouse counterparts (assuming they exist) will also bind to some primate cells. An example is the antigen recognized by the monoclonal antibody 12E7 (Levy *et al.*, 1979), which is present on human, chimpanzee and gorilla fibroblasts, but not on orang-utang, muntjac, rabbit or rodent cells (Goodfellow, 1983; and unpublished results). Thus, the 12E7 antigenic determinant is species-restricted rather than truly species-specific. However, its absence from mouse cells allowed assignment of the controlling locus initially to human Xp (Goodfellow *et al.*, 1980), and subsequently to Xp22.3 → Xpter (Goodfellow *et al.*, 1983). An analogous gene was also found on the human Y chromosome (Goodfellow *et al.*, 1983 and see Section 3.5.2). Further investigation of the genetics of these loci in primates and other mammals might prove useful in studies of sex chromosome evolution.

Polyclonal, heterospecific antisera have proved of value in human gene mapping by somatic cell genetics (Section 3.4.1; Table 3.1) and also in other genetic systems: e.g. the use of alloantisera against HLA region (see Chapter 4 and Parham and Strominger, 1982) and red blood cell (Chapter 2) surface determinants in man, and against lymphocyte determinants in mouse (see Chapter 1 and McKenzie and Potter, 1979). However, there can be numerous problems with such antisera, some of which are: the presence of contaminating antibodies, low titre, paucity of material and non-reproducibility. In this

Table 3.1 Genes for cell surface antigens characterized by polyclonal antisera

Chromosomal location	Antigen	Reference
6	HLA (S5)	a
7	EGFR (S6)	b
11	S1, S2 S3, S4	c,d,e,f
12	S8	g
15	β_2m	h
17	S9	i
21	Interferon receptor (S14)	j
X	S10 (SAX)	k

* References: (a) van Someren *et al.* (1974); (b) Aden and Knowles (1976); Carlin and Knowles (1982); (c,d) Jones *et al.* (1975); (e) Jones and Puck (1977); (f) Buck and Bodmer (1974); (g) Seravalli *et al.* (1978); (h) Goodfellow *et al.* (1975); (i) Cicurel and Croce (1977); (j) Chan *et al.* (1979); (k) Buck and Bodmer (1976).

respect, monoclonal antibodies are superior reagents in that they are monospecific, can be obtained with high titres and are available in theoretically unlimited quantities. Monoclonal antibodies are therefore the reagents of choice.

3.4 THE USE OF ANTIBODIES IN SOMATIC CELL GENETICS

3.4.1 Initial mapping studies using polyclonal antibodies

When the first human–mouse somatic cell hybrid cell lines were made in 1967 by Weiss and Green, it was noted that antisera made in rabbits against human fibroblasts, after adsorption with mouse cells, would promote agglutination of hybrid cells and human erythrocytes. The extent of this activity was dependent upon the number of metacentric chromosomes, most of which would be human, in different hybrids. Thus, human antigens are present on the hybrid cell surface. This was confirmed later by others (Nabholz *et al.*, 1969; Kano *et al.*, 1969), and the Stanford group suggested that genes for human antigens might be segregating in hybrids together with one of the lactate dehydrogenase

(LDH) loci. Indeed, this was shown to be the case by Puck and colleagues (Puck *et al.*, 1971) who demonstrated that genes for human antigens and LDH_A were on the same human chromosome (i.e. showed synteny). These antigens were called A_L, for 'lethal' antigens, since the method of assay was cytotoxicity (Oda and Puck, 1961). In this assay, cells are plated at low dilutions in the presence of antiserum and complement. Those cells with A_L (in this case) are lysed by the antibody–complement system, whereas those without A_L survive to form colonies. In the early seventies, methods for identifying each of the human chromosomes became available (Caspersson *et al.*, 1971; Seabright, 1972; Bobrow and Cross, 1974). This allowed identification of human chromosomes in hybrids, and subsequently syntenic loci (Santachiara *et al.*, 1970; Puck *et al.*, 1971) could be assigned to particular chromosomes. Thus, when *LDHA* was assigned to chromosome 11 (Boone *et al.*, 1972), the A_L loci were simultaneously placed on that chromosome. By the use of selected mutants and antisera from a range of animals immunized with several human or hybrid cell types, it was possible to split the A_L system into at least three distinct antigenic identities (Jones *et al.*, 1975; Jones and Puck, 1977). These were named a_1, a_2 and a_3, and by using hybrids with deletions in chromosome 11, it was possible to assign a_1 and a_3 to 11p13→11pter and a_2 to 11q13→11qter (Kao *et al.*, 1977). The A_L antigens were later renamed S1, S2 and S3.

Using the immunization protocol outlined in the previous section, Buck and Bodmer (1974) were also able to produce conventional antisera to antigens coded by chromosome 11. They used the genetic isolation procedure outlined in Section 3.1 and immunized mice with a hybrid containing only human chromosomes 11, 13 and X, and after adsorption with mouse cells, produced an antiserum whose cytotoxic activity against hybrids segregated with chromosome 11. It was later shown (Buck *et al.*, 1976) that only the short arm of chromosome 11 was necessary for antigen expression. This antigen (or antigens) was called SA-1 (species antigen 1) and subsequently renamed S4, and may be identical, or related to S1 of the A_L system, since both of these antigens have been shown to be on glycolipids (Jones *et al.*, 1979; D. Marcus, unpublished results). It is interesting that when whole human cells, or hybrids with more than one chromosome, are used as immunogens, the dominant immune stimulus is provided by antigens encoded by chromosome 11. Evidently, this is not simply due to all or most cell surface determinants being coded on this chromosome as Table 3.1 demonstrates. This gives a list of surface determinants mapped by techniques involving polyclonal antibodies: at least a third of the human chromosomes harbour genes controlling surface antigens mapped by these methods. Four of these antigens have partially characterized functions: HLA, β_2-microglobulin (β_2m), EGF receptor (EGFR) and interferon receptor. Possible biological functions of the other surface molecules identified genetically remain to be determined and it is possible that monoclonal antibodies to these determinants will be valuable in this respect.

3.4.2 Mapping with monoclonal antibodies

The genetic use of monoclonal antibodies against cell surface molecules was first demonstrated by Barnstable *et al.* (1978), who immunized mice with human tonsil leucocytes and isolated several useful antibodies: W6/1, specific for blood group A erythrocytes; W6/32, anti-HLA−A, B, C heavy chain; three monoclonal antibodies whose binding correlates with the presence of human chromosome 11 in hybrids, including one W6/34, which was cloned and shown to recognize an antigen on a glycolipid. The gene controlling W6/34 antigen expression was further mapped to the short arm of chromosome 11.

The use of monoclonal antibodies has been exploited further such that now antigenic markers for approximately half the human chromosomes are available (Table 3.2), although not all of these are expressed on all human cell types. An example of the mapping techniques is shown for the transferrin

Table 3.2 Genes for cell surface antigens identified by monoclonal antibodies

Chromosomal Location	Gene	Antibody	Antigen	Reference†
3	*TFRC*	OKT9	Transferrin receptor	a
4	—	OKT10	45 K Protein	b
6	*HLA−A, B, C*	W6/32	HLA−A, B, C	c
7	*EGFR*	EGFR1	Epidermal growth factor receptor	d
11	*MIC1**	W6/34	Glycolipid	c
11	*MIC4*	F10.44.2	105 K Glycoprotein	e
11	*MIC9*	4D12	100 K Protein	f
11	*MIC8*	TRA1.10	80 K, 40 K Protein	f
11	*MIC11*	163A5	200 K, 150 K, Protein	g
12	*MIC3*	602	21 K Protein	h
15	*B2M*	BBM1	β_2-Microglobulin	i
15	*MIC7*	28.3.7	95 K Protein	j
15	*MIC12*	302	—	k
17	*MIC6*	H207	125 K Protein	l
X	*MIC2X*	12E7	34 K Protein	m
X	*MIC5*	RI	—	n
Y	*MIC2Y*	12E7	34 K Protein	o

*The designation *MIC* is a provisional local name for genes mapped by monoclonal antibodies at ICRF, a nomenclature agreed upon at the 6th International Human Gene Mapping Workshop, Oslo, 1981.

†References: a Goodfellow *et al.* (1982a); b Katz *et al.* (1983); c Barnstable *et al.* (1978); d Goodfellow *et al.* (1981); Waterfield *et al.* (1982); e Goodfellow *et al.* (1982b); f Tunnacliffe *et al.* (1983); g Woodruffe *et al.* unpublished; h Andrews *et al.* (1981); i Brodsky *et al.* (1979); j Blaineau *et al.* submitted; k Walsh *et al.* unpublished; l Bai *et al.* (1982); m Goodfellow *et al.* (1980); n Hope *et al.* (1982); o Goodfellow *et al.* (1983).

receptor (TFR) in Tables 3.3 and 3.4 (Goodfellow *et al.*, 1982a). The antibody used against TFR was OKT9 (Reinherz *et al.*, 1980), which was proved to recognize the receptor after co-precipitation by the antibody of labelled transferrin with TFR (Sutherland *et al.*, 1981). For mapping the *TFR* gene, initially a panel of independent hybrids was tested (Table 3.3) by indirect radioimmunoassay (IRIA; Williams, 1977; Footnote to Table 3.3), followed by confirmation with a set of secondary subclones of a primary hybrid which segregates the

Table 3.3 Testing a panel of independent hybrids for OKT9 antigen

Cell*	Human chromosomes present	$10^{-3}\times$ Radioactivity bound (cpm)	
		OKT9†	P3.X63.Ag8 (negative control)
MOG7	1,3,4,5,7,8,10,11,12,13,15,16,18,21,X	13.8	1.8
SIR2	1,2,3,4,5,6,7,8,10,11,12,13,14,15,16,17, 18,19,20,21,22,X	11.2	0.7
HORL4.1.1.B6	1,3,10,11,13,15,18,X	8.9	0.5
MOG13/17	3,21,22,X	7.3	0.4
DUR4.3	3,5,10,11,12,13,14,15,(17),18,20,21,22,X	3.9	0.5
DUR5	3,5,10,11,12,13,14,15,17,18,20,21,22,X	5.1	1.5
3W4C1.5	7,10,11,12,14,15,21,X	2.3	1.3
HORL9D2	11,15,X	1.8	1.5
HORP9.5	10,11,12,14,21,X	1.3	0.6
THYB.133	21,X	0.7	0.4
F4SC13.CL12	1,9,14,X	2.0	1.0
4W10.R3	8,20,21	2.2	1.1
CL21	7	2.7	1.9
FIR5R3	14,18	1.1	0.7
MOLT-4	Human T cell line	12.1	0.5
G3.32.2	Burkitt's lymphoma line	22.2	0.6
HFL121	Human fibroblast	6.4	0.4
1R	Mouse L cell	2.1	1.5
3T3	Mouse cell (fibroblast-like)	1.6	1.5
RAG	Mouse adenocarcinoma	1.8	1.6
BW5147	Mouse thymoma	1.6	1.0

*References to hybrids and cell lines are given in Goodfellow *et al.* (1982a).

†2×10^5 attached or 5×10^5 suspension cells are incubated for 1 h at room temperature with saturating titres of first antibody, washed three times and then incubated with 2×10^5 cpm of iodinated rabbit anti-(mouse IgG for a further hour at 4°C. Cells are washed four times and counted. Values greater than three times background (with P3.X63.Ag8) are taken as positive and are underlined.

Table 3.4 Testing subclones of DUR4 and MOG13 for OKT9 antigen

Hybrid	Human chromosomes	$10^{-3}\times$ Radioactivity bound (cpm)	
		OKT9	P3.X63.Ag8
DUR4.3*	*3*,5,10,11,12,13,14,15,17,18,20,21,22,X	3.9	0.5
DUR4.4*	*3*,11,12,13,14,15,17,18,21,22,X	5.1	0.4
DUR4.5*	*3*,5,10,11,12,13,14,15,17,18,20,21,22,X	4.4	0.7
DUR4R.1	5,11,12,13,14,17,18,20,21,22	1.2	0.8
DUR4R.3	*3*,5,11,12,13,14,17,18,20,21,22	7.5	1.2
DUR4R.4	*3*,5,11,12,13,14,17,21,22	10.3	1.0
MOG13.9	X	0.9	0.7
MOG13.10	1,*3*,21,22,X,Y	7.3	0.4
MOG13.17	*3*,21,22,X	8.4	0.6
MOG13.22	1,21,22,X	0.9	0.5

*Contains an active X→15 translocation.

chromosome in question (Table 3.4). It can be seen by inspection that only human chromosome 3 segregates with OKT9-binding activity in both the hybrid panel and the two sets of hybrid subclones. This assignment has been confirmed independently by Enns *et al.* (1982) using different techniques.

The assignment of *EGFR* to chromosome 7 using a monoclonal antibody (Table 3.2) confirms a previous result (Shimizu *et al.*, 1980; Davies *et al.*, 1980), where EGF binding to hybrids was measured, the rodent parent of which lacked receptor. It was also shown recently (Carlin and Knowles, 1982) that the 165 K marker for chromosome 7 identified with polyclonal antisera (Aden and Knowles, 1976) is also EGFR. Thus, the experiments with the monoclonal antibody have added weight to the assignments made using polyclonal antibodies and labelled ligand binding.

Table 3.2 shows that at least five different genes coding surface antigens are present on chromosome 11. Their possible relationship to genes identified using polyclonal antisera has been discussed previously (Tunnacliffe *et al.*, 1983b): it seems likely that $A_L a_1$ is identical, or closely related to W6/34 antigen. There may also be a relationship between $A_L a_3$ and F10.44.2 and 163A5 antigens, and between TRA1.10 and 4D12 antigens and $A_L a_2$. It has also been shown recently that the monoclonal antibody 4F2 recognizes an antigen mapping to 11 (Peters *et al.*, 1982). This may be binding the same molecule(s) as TRA1.10, since both antibodies precipitate bands at 80 K and 40 K on SDS/PAGE.

The question arises as to why so many genes controlling cell surface antigens exist on chromosome 11. It is possible that there are only one or a few genes responsible which code for glycosyltransferases which create novel (to the mouse) antigenic determinants on mouse surface molecules. However, the five antibody-binding activities are genetically separable using hybrids with differing fragments of chromosome 11 present and also show differing distributions on cell lines and tissues (Tunnacliffe *et al.*, 1983a). In the case of W6/34, the antigenic determinant definitely resides on a glycolipid which suggests a human glycosyltransferase as the active agent in creating the epitope. Indeed, if W6/34 antigen is identical to S1 (A_La_1) as it seems to be, then mouse genes in at least three complementation groups are necessary for its expression (Jones *et al.*, 1979). This molecule is thus probably found on hybrid cell surfaces as a result of the action of mouse and human genes in concert. It would be of interest to compare in detail, the human and hybrid antigen-carrying molecules responsible for W6/34 and anti-S1 binding. The other four chromosome 11-specific monoclonal antibodies precipitate proteins, however, and it is likely that at least a part of the antigenic determinants they recognize are due to proteins whose coding sequences are on human chromosome 11.

It is probable that a monoclonal antibody recognizing a marker antigen for each of the human chromosomes will soon be available, in which case such a panel of antibodies will be useful for the rapid characterization of human chromosomes present in hybrids. It is evident that antigenic analysis of hybrids using monoclonal antibodies and IRIA or ELISA can produce a result in a few hours, whereas karyotype or isozyme analysis (where 23 gels need to be run per hybrid) take considerably more time and effort.

A panel of monoclonal antibodies, in conjunction with the fluorescence-activated cell sorter (FACS), can also be used for hybrid manipulation and studies of antigen expression, as described in the next section.

3.5 THE USE OF THE FACS

3.5.1 Quantitative and qualitative analysis of cell surface antigens in hybrid cell populations

The FACS (Loken and Herzenberg, 1975) allows the objective determination of the percentage of a cell population expressing an antibody-defined cell surface determinant, and in the case of a hybrid population, immediately provides data on the percentage of cells containing the marker-defined chromosome. An example is shown in Fig. 3.2. Scatter plots are shown of FACS analysis of a hybrid HORL9D2RM (Goodfellow *et al.*, 1982b). This mass culture was derived from a hybrid HORL9D2 (Goodfellow, 1975) which contains human chromosomes X,11 and 15 against a mouse $HPRT^-$ L-cell background. The X chromosome has been removed by selection in 6-thioguanine (6TG) to give HORL9D2RM. Panel (a) shows the negative control

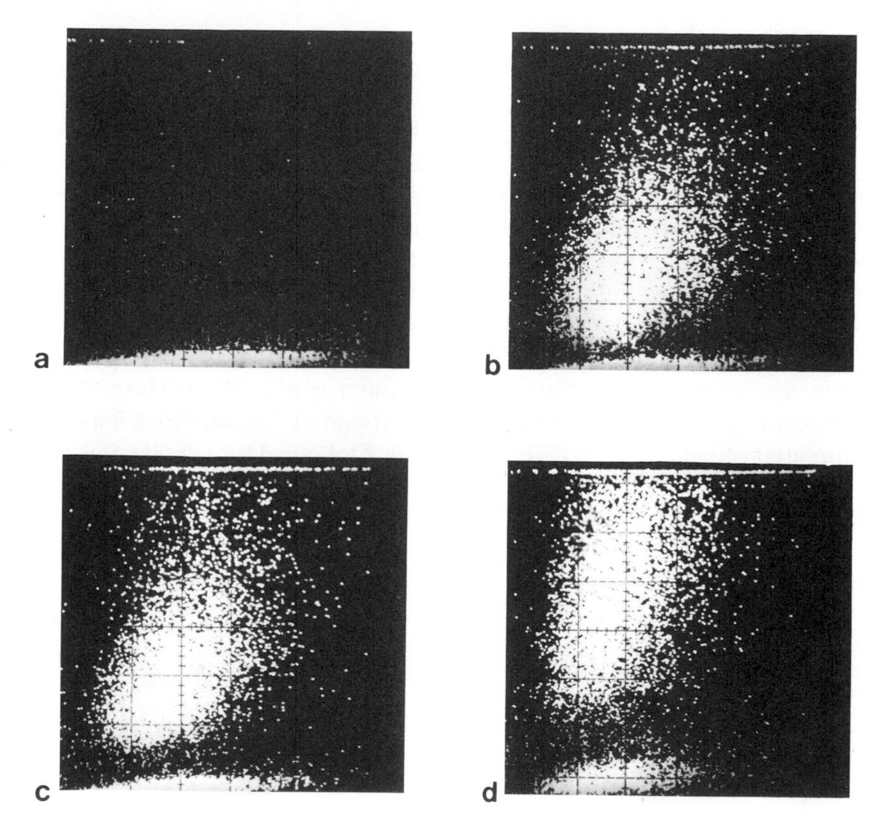

Fig. 3.2 FACS analysis of hybrid HORL9D2RM with monoclonal antibodies (a) P3.X63.Ag8; (b) F10.44.2; (c) W6/34; (d) BBM1.

reaction with mouse myeloma P3.X63.Ag8 (Kohler *et al.*, 1976). Panels (b) and (c) show the scatter plots after indirect immunofluorescence (IIF) with monoclonal antibodies F10.44.2 (Dalchau *et al.*, 1980) and W6/34 (Barnstable *et al.*, 1978) respectively. Incubation of hybrid cells with these antibodies was followed, after washing, by reaction with fluorescein-labelled rabbit anti-mouse IgG to label positive cells. Both F10.44.2 (Goodfellow *et al.*, 1982b) and W6/34 (Barnstable *et al.*, 1978; Goodfellow *et al.*, 1982b) recognize antigens whose expression is controlled by chromosome 11 (Table 3.2). Panel (d) shows staining with BBM1, a monoclonal anti-β_2m (Brodsky *et al.*, 1979), a marker for human chromosome 15 (Goodfellow *et al.*, 1975). It can be seen by the presence of negative populations that not all cells of HORL9D2RM have chromosomes 11 and/or 15. When the hybrid is stained with combinations of two antibodies it is found that both 11-markers together give the same size positive populations as an 11-marker alone, but that staining with an 11- and

15-marker together increases the size of the positive population. This means that some hybrids have only chromosome 11 or 15 and also suggests synteny of the genes encoding W6/34 and F10.44.2 antigens.

An extension of this technique could be used to demonstrate synteny between two antigen-controlling loci. The first antibody might be used to sort a mass culture into two populations, and then each of these populations restained with the second antibody and a different fluorochrome. If the two loci are indeed syntenic, the positive and negative populations from the first sort should be approximately 100% and 0% positive after the second staining (assuming the second fluorescent stain is distinguishable from the first).

It is possible to demonstrate the co-existence of two antigens in a population by the use of two-colour fluorescence (Loken *et al.*, 1977). Here, stepwise incubations and stainings are carried out: antibody 1 followed by a fluorescein-conjugated second antibody, then antibody 2 followed by a rhodamine-linked

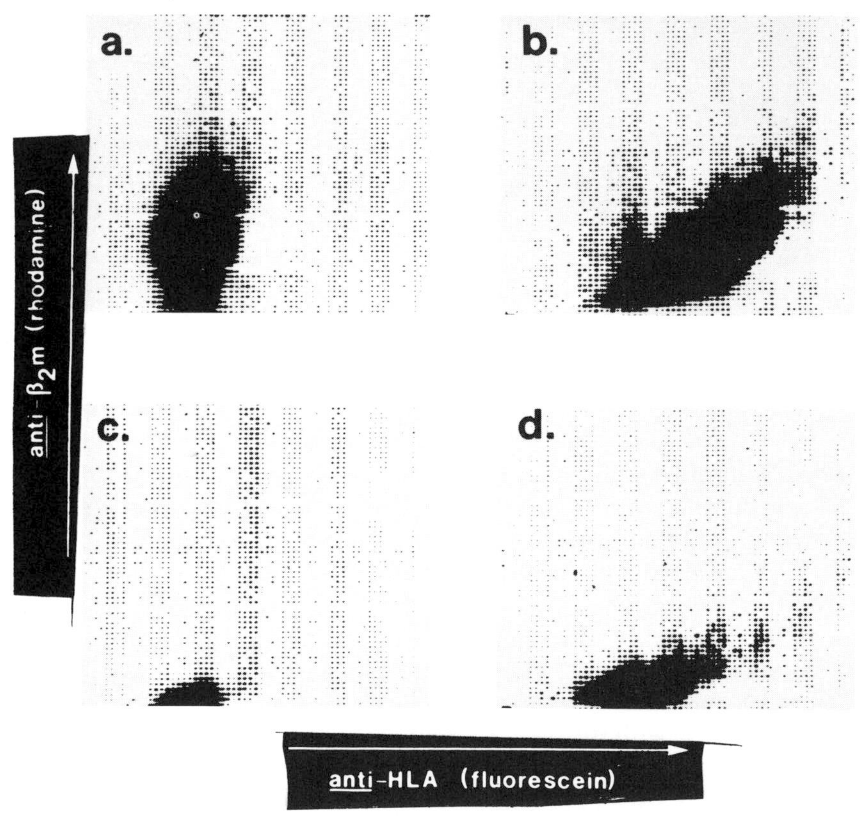

Fig. 3.3 Two-colour FACS analysis of β_2-microglobulin and HLA on subpopulations of hybrid FRY4 which are (a) HLA^- β_2m^+; (b) HLA^+ β_2m^+; (c) HLA^- β_2m^- and (d) HLA^+ β_2m^-. Taken from Kamarck *et al.* (1982) with permission.

second reagent. This technique has been used by Kamarck *et al.* (1982) to examine the co-expression of HLA and human β_2m on a hybrid cell surface. HLA was indirectly stained with fluorescein and β_2m with rhodamine and then both fluorochromes were simultaneously excited. After filtering out common fluorescent wavelengths, it is possible to identify separate signals from each fluorescent stain and hence from HLA and β_2m. The authors examined four populations which had previously been sorted from a hybrid FRY4 by one-colour fluorescence, for the various combinations of HLA and human β_2m, namely (a) $HLA^- \beta_2m^+$, (b) $HLA^+ \beta_2m^+$, (c) $HLA^- \beta_2m^-$ and (d) $HLA^+ \beta_2m^-$. Two-colour staining confirmed the one-colour results, as shown in Fig. 3.3, where the panels correspond to the phenotypes given above. It is also possible to two-colour sort for a subpopulation with any of the four combinations. Kamarck *et al.*, using FACS technology, were also able to show that the presence of human β_2m increased the level of expression of HLA and mouse H-2 antigens on the hybrid cell surface. Two-colour analysis could also be used to show synteny of genes controlling cell-surface determinants: syntenic genes should give only ++ and −− populations.

Synteny in a single population between the loci for a surface antigen and an enzyme can be shown by sorting on the FACS: positive and negative populations are obtained from a hybrid segregating an antigenic marker, cell extracts made and enzyme assays carried out. This has been used to show synteny between F10.44.2 and W6/34 antigens, and LDHA (EC1.1.1.27), and between β_2m and pyruvate kinase (PK; EC2.7.1.40) (Table 3.5; Goodfellow *et al.*, 1982b). Sorting for a particular antigen, followed by DNA extraction from subsequent populations should also enable synteny tests for a cloned DNA fragment to be performed. These techniques are considerably more rapid than

Table 3.5 FACS sorting followed by isozyme analysis of HORL9D2RM

		Isozyme testing*	
Selecting antibody	Antigen population	LDHA	PK
F10.44.2	Positive	++	++
	Negative	−	+
W6/34	Positive	++	++
	Negative	−	++
BBM1	Positive	++	++
	Negative	++	−

*++, Strong expression of human isozyme; +, moderate expression of human isozyme; −, no expression of human isozyme.

conventional subcloning techniques and individual analysis of resultant clones, which were mandatory before FACS technology was available.

3.5.2 Hybrid manipulation

Antibodies to chromosome-specific cell surface antigens are of use not only for quantifying the human chromosome contribution to a somatic cell hybrid population and for synteny tests, but also for altering the human contribution. Polyclonal antibodies to HLA, β_2m and SA-1 have previously been used to select against hybrids containing chromosome 6, 15 and 11 by cytotoxicity in the presence of complement (Jones *et al.*, 1976; Goodfellow, 1975). Thus, in a hybrid population which is segregating a particular non-selected human chromosome, only those cells lacking the chromosome, and consequently the respective antigen, will survive. Hybrid manipulation in this way is essentially a negative selection system and has largely been superseded by the use of the FACS. The FACS allows selection of populations which are either positive or negative for a surface antigen when used in conjunction with the appropriate antibody. Dorman *et al.* (1978) used a conventional antiserum against the X-linked antigen SAX (Buck and Bodmer, 1976) to successfully sort a hybrid mass culture into two viable populations, positive and negative for the X chromosome. These initial experiments showed the potential of the approach: current work has employed monoclonal antibodies in place of heterospecific antisera.

Cell sorting followed by cloning of the resultant populations can lead to directed and stable hybrid manipulation: subclones with or without a particular chromosome can be isolated. Following the example of HORL9D2RM again, it was possible to isolate '15-only' hybrids by selecting a population negative for W6/34 from which clones were derived. As assayed by IRIA and isozyme testing, these had lost chromosome 11 markers, but retained markers for chromosome 15 (Goodfellow *et al.*, 1982b; and unpublished results). We have used these techniques to demonstrate the Y-linkage of a gene controlling expression of the 12E7 antigen (Levy *et al.*, 1979). A hybrid, AMIR2, made between fibroblasts from a male with X-linked icthyosis (46; t(X:Y;Xqter→Xp22.3:Yqter→Yp1.1);Y) and mouse HPRT$^-$ L cells, was shown to be 12E7-positive by IRIA. When examined on the FACS, however, only 24% of cells were positive (Goodfellow *et al.*, 1983). Sorting for a 12E7-negative population, followed by karyotype analysis, showed that these cells had lost the normal Y chromosome, present in the original hybrid, although still retaining the X→Y translocation chromosome in 94% of cells. This simple experiment provides several pieces of information. (a) 12E7 expression is coincident with the presence of the normal Y chromosome in AMIR2; (b) although a 12E7-controlling gene had previously been assigned to Xp (Goodfellow *et al.*, 1980), the presence of the region Xqter→Xp22.3 in hybrids was not sufficient for 12E7 expression. This suggests that the X-linked locus, *MIC2*, controlling antigen expression, is in the region Xp22.3→Xpter. (c) By the same logic, the Y locus must be in the region Yq1.1→Ypter, which is not included in the translocation chromosome.

Other data from different hybrids are concordant with the results of this experiment (Goodfellow *et al.*, 1983) and present an intriguing picture of two similar, if not identical, structural genes on the X and Y chromosomes. Furthermore, the regional mapping places these genes in the regions which pair with each other during meiosis (Pearson and Bobrow, 1970), and *MIC2X* is in the region of the X chromosome thought not to be inactivated in females (Race and Sanger, 1975; Shapiro *et al.*, 1979; Muller *et al.*, 1980). The identification of potentially homologous functional genes on the X and Y chromosomes may thus have important implications for sex chromosome evolution, X-inactivation and sex determination.

3.5.3 Other uses of the FACS

The FACS is also proving useful in other areas, for example, for identifying the products of cloned genes: Barbosa *et al.* (1982) identified cosmid clones for HLA-A2 and HLA-B7 after transfecting those clones from a genomic library which hybridized to an HLA cDNA probe into L cells, and monitoring HLA expression with allospecific monoclonal antibodies. There is also a report (Stanners *et al.*, 1981) of the use of the FACS to isolate by sorting of a population of rodent cells transfected with a hybrid genomic library, a clone responsible for the expression of a cell surface antigen associated with human chronic lymphocytic leukaemia, and inducing a more malignant phenotype in hybrids and transfectants. The authors judge the gene to be present in 100–1000 copies in the hybrid genome, and this technology may prove more difficult for the cloning of low copy number genes. Recently, however, Kavathas and Herzenberg (1983) transfected mouse L cells with human DNA and isolated transfectants positive for HLA, β_2-microglobulin, Leu-1 (OKT1) and Leu-2 (OKT5/8) by FACS sorting. This should allow identification of the specific human sequences responsible for the Leu-1 and Leu-2 antigens after further rounds of transfection and sorting.

The use of the FACS is not limited to the analysis of cell surface determinants only: we have used a monoclonal antibody to a cytokeratin to examine, in fixed, permeablized cells, the differential expression of this protein in differentiating human–mouse teratocarcinoma hybrids (Benham *et al.*, 1983). Expression of SV40 T-antigen in SV40-transformed human cells and somatic cell hybrids has also been investigated with these techniques. It has also been possible to demonstrate SV40 large T in an SSEA-1-positive population of embryonal carcinoma hybrids after assaying a FACS-sorted cell extract (unpublished results).

3.6 CONCLUSIONS AND PROSPECTS

We have reviewed briefly the contribution of the use of antibodies to somatic cell genetic analysis, and in particular noted the progression from the use of polyclonal to monoclonal antibodies, which parallels a general trend in the use of antibodies as analytical tools in cellular and molecular biology. In the near-future, there is a high probability that monoclonal antibodies will be available which recognize at least one marker antigen for each human chromosome. The current picture, summarized in Table 3.2, indicates that, to date, almost half of the human chromosomes are covered. A complete panel will allow rapid analysis of the human chromosome content of hybrids and will further potentiate the use of the FACS in mapping and expression studies using hybrids. The use of antibodies for gene mapping has meant that species-restricted functional gene assays are not required. Instead, structural and immunological properties of the gene product are exploited; further use of this approach should lead to the identification of new genetic systems. An example is provided by the 12E7 antigen-controlling locus, *MIC2*: gene mapping has shown the existence of two related forms of *MIC2*, one on the tip of the short arm of the X chromosome, probably not X-inactivated and having some relationship with the *Xg* locus (Goodfellow and Tippett, 1981), and the other in the euchromatic region of the Y chromosome (Goodfellow *et al.*, 1983). This is the first demonstration of a structural gene on the human Y chromosome. Many other gene products identified by monoclonal antibodies and listed in Table 3.2 have no known function as yet. Possibly, some of these molecules have already been identified by other methods, but others may represent novel identities which can be explored further for functions.

Most of the monoclonal antibodies in Table 3.2 have a general tissue distribution, but it is also possible to investigate the genetics of specialized functions by use of the appropriate hybrid parents. Also, many monoclonal antibodies against haematopoietic cells have been produced (see Foon *et al.*, 1982; Reinherz and Schlossman, 1980 for reviews) and some of the genes controlling cognate antigen expression are now being mapped, for example, OKT10 antigen is encoded by human chromosome 4 (Katz *et al.*, 1983).

The antibody/hybrid approach is not, of course, restricted to the analysis of surface components. For example, complement component C3, secreted by fibroblast hybrids, has been mapped to human chromosome 19 using a monoclonal antibody (Whitehead *et al.*, 1982). It is also possible to map enzyme loci using antibodies: Vora *et al.* (1982) used a monoclonal antibody to human muscle-type phosphofructokinase (PFKM) to map the *PFKM* gene to chromosome 1. The conventional approach of isozyme separation by gel electrophoresis was not possible in this case due to at least three structural loci, in both humans and rodents, expressing different forms of PFK activity. This would give an extremely complex gel pattern, whereas the monoclonal antibody could be used to precipitate enzyme from hybrid cell extracts.

As the number of gene assignments increases, it is possible to ask questions about the overall organization of the human genome. One question pertinent to the work described here is: are genes for cell surface determinants clustered? This question should be refined slightly to emphasize the structural genes of surface proteins, since one possible complication of the antigenic recognition of cell surface molecules is that the determinant may be produced by the addition of carbohydrate to a nascent protein chain. In this case, if the addition is species-specific, or, in combination with a protein structure, produces a species-specific determinant, it may be the gene for a glycosyltransferase which is mapped. In the case of a surface glycolipid (such as W6/34 antigen), this is almost certainly what is happening. A corollary of this is that one glycosyltransferase may produce several different epitopes on different recipient molecules. To guard against this, it may be necessary to take the approach of Whitehead *et al.* (1982) where it was demonstrated that the monoclonal antibody recognizing human C3 also precipitated a non-glycosylated form of the molecule from poly-A^+ RNA extracts translated *in vitro*. Having refined the question in this way, we can immediately answer that no one chromosome carries all the surface protein genes, but that these appear to be spread over a number of chromosomes. This is expected from analysis of other known gene assignments. For example, whereas in *E. coli*, genes of a particular biochemical pathway are often contiguous and expressed in the order dictated by the biochemistry, no such organization is found in the human genome. Indeed, functionally related genes are scattered over several or many chromosomes, although four enzymes of the glycolytic pathway map to chromosome 1.

Neither are evolutionarily related genes necessarily linked: genes for the subunits of the same enzyme can be found on different chromosomes (e.g. *LDHA* on 11, *LDHB* on 12), as can other related structural loci (e.g. β-globin cluster on 11, α-globin cluster on 16). A similar pattern is seen for surface protein genes: thus, while sequence-related HLA and DR genes are linked on chromosome 6, β_2-microglobulin (15) and immunoglobulin (2, 14, 22) genes, whilst having some sequence relationships, are clearly not linked. However, both polyclonal and monoclonal antibodies have identified a number of antigens encoded by chromosome 11 which appear to be on distinct molecules and which are genetically separable. It is known that one of these antigens is present on a glycolipid, although the others are defined by antibodies which precipitate proteins. However, the possibility remains that genes for a small number of glycosyltransferases are being mapped and further experiments will be necessary to clarify this point. In the mouse, it is now emerging that a large cluster of surface antigen genes is grouped around the β_2m locus on chromosome 2 (Meruelo *et al.*, 1982; see also Chapter 1). In humans, β_2m is on chromosome 15 (Goodfellow *et al.*, 1975) and recently we have used monoclonal antibodies to map two further genes for surface determinants to this chromosome (Table 3.2; Blaineau *et al.*, in press; unpublished work). In addition, Sakaguchi and Shaws (1982) have mapped the gene conferring coronavirus 229E susceptibility, probably surface-mediated, to chromosome 15. This makes at least four determinants encoded by this chromosome and it may be that a comparable picture to that of the mouse is forming for this locus. The next few years should answer the question of clustering, and possible relationships of members of clusters to each other in terms of evolution, structure and function.

A far-reaching consequence of gene mapping in general is the possible application to medical genetics. The most dramatic example is that of the oncogenes, genes implicated in cell transformation, approximately fifteen of which have been mapped in the last two years to particular chromosomes and chromosomal regions. It has been found in most cases that these regions are involved in rearrangements or deletions in specific cancers, which may reflect a role of oncogenes in the aetiology of these diseases (reviewed in Rowley, 1983). If it is not possible to determine the presence of a genetic defect directly or when the nature of the defect is not known (e.g. muscular dystrophy), it may be possible to identify a genetic polymorphism or trait which is linked to the disease locus. A classical example is that of haemophilia and its linkage to *G6PD* on the X chromosome: if a mother carrying the haemophilia defect is heterozygous for *G6PD*, family studies will show which allele is associated with the disease locus, and by determining the allelic form of G6PD (glucose 6-phosphate dehydrogenase) in the mother's male offspring *in utero*, it can be determined whether the child might be affected. Genetic counselling along these lines could help to eliminate rapidly unwanted genes from the population if such pregnancies were terminated. An important consideration in such cases is the genetic distance between loci: the larger this distance, the greater the likelihood of cross-over between loci, which would give an incorrect evaluation of whether a child will carry a genetic defect when the above methods were followed. To date, the best-characterized heritable condition involving a defect in a cell surface molecule is familial hypercholesterolaemia. This is a condition associated with defective low-density lipoprotein receptors (LDLR). There is reported to be weak linkage between *C3* and *LDLR* (Ott *et al.*, 1974; Berg and Heiberg, 1976; Elston *et al.*, 1976), which would place *LDLR* on chromosome 19. However, more recently, using hybrids between human fibroblasts and hamster cells lacking functional LDLRs, one of these groups has suggested chromosomes 5 or 21 or both as being involved in LDLR activity in hybrids (Maartmann-Moe *et al.*, 1982). The situation is, therefore, not clear at present and the use of available monoclonal antibodies (Beisiegel *et al.*, 1981) may solve the problem. Once a chromosomal assignment is made, it might be possible to show linkage in family studies with a polymorphic locus which would be useful for genetic counselling.

Conversely, it might be possible with genetic diseases of unknown cause, but where the disease locus has been mapped, to inspect the human gene map for linked or syntenic genes, and then to assay these functions in affected individuals for a relationship. Negative mapping data may also be useful, at least in eliminating certain candidates for the gene defect. For example, haemochromatosis, an iron-storage deficiency, is associated with a defect on chromosome 6. This suggests that the condition is not due to a defect of the transferrin receptor, whose gene maps to chromosome 3. A defect in the ferritin structural gene is also ruled out, since this gene maps to chromosome 19 (Caskey *et al.*, 1983). However, linkage of the defect to *HLA-A* has been shown (Simon *et al.*, 1980; Edwards *et al.*, 1980), and this allows pre-morbid identification of patients, who should then be protected from iron overloading. An expansion of the human gene map by methods including those reviewed here should allow some of these suggestions to be realized and further possibilities explored.

REFERENCES

Aden, D.P. and Knowles, B.B. (1976), *Immunogenetics,* **3,** 209–221.

Andrews, P.W., Knowles, B.B. and Goodfellow, P.N. (1981), *Somat. Cell Genet.,* **7,** 435–443.

Bai, Y., Sheer, D., Hiorns, L., Knowles, R.W. and Tunnacliffe A. (1982), *Ann. Hum. Genet.,* **46,** 337–347.

Barbosa, J.A., Kamarck, M.E., Biro, P.A., Weissman, S.M. and Ruddle, F.M. (1982), *Proc. Natl. Acad. Sci. U.S.A.,* **79,** 6327–6331.

Barnstable, C.J., Bodmer, W.F., Brown, G., Galfre, G., Milstein, C., Williams, A.F. and Zeigler, A. (1978), *Cell,* **14,** 9–20.

Bastin, J.M., Kirkley, J. and McMichael, A.J. (1982), in *Monoclonal Antibodies in Clinical Medicine.* Academic Press, London.

Beisiegel, U., Schneider, W.J., Goldstein, J.L., Anderson, R.G.W. and Brown, M.S. (1981), *J. Biol. Chem.,* **256,** 11923–11931.

Benham, F.J., Wiles, M.V., Banting, G., Andrews, P. and Goodfellow, P.N. (1983), in *Human Teratomas* (I. Damjanov, B.B. Knowles and D. Solter, eds), Humana Press, New Jersey.

Berg, K. and Heiberg, A. (1976), *Cytogenet. Cell Genet.,* **16,** 266–270.

Bobrow, M. and Cross, J. (1974), *Nature (London),* **251,** 77–79.

Boone, C., Chen, T.R. and Ruddle, F.H. (1972), *Proc. Natl. Acad. Sci. U.S.A.,* **69,** 510–514.

Brodsky, F.M., Bodmer, W.F. and Parham, P. (1979), *Eur. J. Immunol.,* **9,** 536–545.

Buck, D.W. and Bodmer, W.F. (1974), *Cytogenet. Cell Genet.,* **14,** 257–259.

Buck, D.W. and Bodmer, W.F. (1976), *Cytogenet. Cell Genet.,* **16,** 376–377.

Buck, D.W., Bodmer, W.F., Bobrow, M. and Francke, U. (1976), *Cytogenet. Cell Genet.,* **16,** 97–98.

Carlin, C.R. and Knowles, B.B. (1982), *Proc. Natl. Acad. Sci. U.S.A.,* **79,** 5026–5030.

Caskey, J.H., Jones, C., Miller, Y.E. and Seligman, D.A. (1983), *Proc. Natl. Acad. Sci. U.S.A.,* **80,** 482–486.

Caspersson, T., Lomakka, G. and Zech, L. (1971), *Hereditas,* **67,** 89–102.

Chan, M.M., Kano, K., Dorman, B., Ruddle, F.H. and Milgrom, F. (1979), *Immunogenetics,* **8,** 265–275.

Cicurel, L. and Croce, C.M. (1977), *J. Immunol.,* **118,** 1951–1956.

Dalchau, R., Kirkley, J. and Fabre, J. (1980), *Eur. J. Immunol.,* **10,** 745–749.

Davies, R.L., Grosse, V.A., Kucherlapati, R. and Bothwell, M. (1980), *Proc. Natl. Acad. Sci. U.S.A.,* **77,** 4188–4192.

Dev, V.G. and Tantravahi, R. (1982), in *Techniques in Somatic Cell Genetics* (J.W. Shay, ed.) Plenum Press, New York and London.

Dorman, B.T., Shimizu, N. and Ruddle, F.H. (1978), *Proc. Natl. Acad. Sci. U.S.A.,* **75,** 2363–2367.

Edwards, C.Q., Cartwright, G.E., Skolnick, M.H. and Amos, D.B. (1980), *Hum. Immunol.,* **1,** 19–22.

Elston, R.C., Namboodiri, K.K. Go, R.C.P., Siervogel, R.M. and Glueck, C.J. (1976), *Cytogenet. Cell Genet.,* **16,** 294–297.

Enns, C.A., Suomalainen, H.A. Gebhardt, S.E., Schroder, J. and Sussman, H.H. (1982), *Proc. Natl. Acad. Sci. U.S.A.,* **79,** 3241–3245.

Foon, K.A., Schroff, R.W. and Gale, R.P. (1982), *Blood,* **60,** 1–19.

Goodfellow, P.N. (1975), D. Phil. Thesis, University of Oxford.

Goodfellow, P.N. (1983), *Differentiation,* **23,***(Suppl.)*, 35–39.

Goodfellow, P.N. and Tippett, P. (1981), *Nature (London),* **289,** 404–405.

Goodfellow, P.N., Jones, E.A., van Heyningen, V., Solomon, E., Bobrow, M., Miggiano, V. and Bodmer, W.F. (1975), *Nature (London),* **254,** 267–269.

Goodfellow, P.N., Banting, G., Levy, R., Povey, S. and McMichael, A. (1980), *Somat. Cell Genet.,* **6,** 777–787.

Goodfellow, P.N., Banting, G., Waterfield, M. and Ozanne, B. (1981), *Cytogenet. Cell Genet.,* **32,** 283.

Goodfellow, P.N., Banting, G., Sutherland, R., Greaves, M., Solomon, E. and Povey, S. (1982a), *Somat. Cell Genet.,* **8,** 197–206.

Goodfellow, P.N., Banting, G., Wiles, M.V., Tunnacliffe, A., Solomon, E., Dalchau, R. and Fabre, J.W. (1982b), *Eur. J. Immunol.,* **12,** 659–663.

Goodfellow, P.N., Banting, G., Sheer, D., Ropers, H.H., Caine, A., Ferguson-Smith, M., Povey, S. and Voss, R. (1983), *Nature (London),* **302,** 346–349.

Harris, H. and Hopkinson, D.A. (1976), *Handbook of Enzyme Electrophoresis in Human Genetics,* Suppl. 1977 and 1978. North-Holland Publishing Co., Amsterdam.

Harris, H. and Watkins, J.F. (1965), *Nature (London),* **205,** 640–646.

Hope, R.M., Goodfellow, P.N., Solomon, E. and Bodmer, W.F. (1982), *Cytogenet. Cell Genet.,* **33,** 204–212.

Jones, C. and Puck, T.T. (1977), *Somat. Cell Genet.,* **3,** 407–420.

Jones, C., Wuthier, P. and Puck, T.T. (1975), *Somat. Cell Genet.,* **1,** 235–246.

Jones, C., Moore, E.E. and Lehman, D.W. (1979), *Proc. Natl. Acad. Sci. U.S.A.,* **76,** 6491–6495.

Jones, E.A., Goodfellow, P.N., Kennett, R.H. and Bodmer, W.F. (1976), *Somat. Cell Genet.,* **2,** 483–496.

Kamarck, M.E., Barbosa, J.A. and Ruddle, F.H. (1982), *Somat. Cell Genet.*, **8,** 385–402.

Kano, K., Baranska, W., Knowles, B.B., Koprowski, H. and Milgrom, F. (1969), *J. Immunol.*, **103,** 1050–1060.

Kao, F.T., Jones, C. and Puck, T.T. (1977), *Somat. Cell Genet.*, **3,** 421–429.

Katz, F., Povey, S., Schneider, C., Sutherland, R., Stanley, K. and Greaves, M. (1983), *Eur. J. Immunol.* (in press).

Kavathas, P. and Herzenberg, L.A. (1983), *Proc. Natl. Acad. Sci. U.S.A.*, **80,** 524–528.

Kennett, R.H., KcKearn, T.J. and Bechtol, K.B. (1981), *Monoclonal Antibodies. Hybridomas: A New Dimension in Biological Analyses.* Plenum Press, New York and London.

Kielty, C.M., Povey, S. and Hopkinson, D.A. (1982a), *Ann. Hum. Genet.*, **46,** 135–143.

Kielty, C.M., Povey, S. and Hopkinson, D.A. (1982b), *Ann. Hum. Genet.*, **46,** 307–327.

Kohler, G. and Milstein, C. (1975), *Nature (London)*, **256,** 495–497.

Kohler, G., Howe, S.C. and Milstein, C. (1976), *Eur. J. Immunol.*, **6,** 292–295.

Kucherlapati, R.S., Baker, R.M. and Ruddle, F.H. (1975), *Cytogenet. Cell Genet.*, **14,** 192–193.

Kull, F.C., Jr., Jacobs, S., Su, Y.-F. and Cuatrecasas, P. (1982), *Biochem. Biophys. Res. Commun.*, **106,** 1019–1026.

Levy, R., Dilley, J., Fox, R.I. and Warnke, R. (1979), *Proc. Natl. Acad. Sci. U.S.A.*, **76,** 6552–6556.

Littlefield, J.W. (1964), *Science*, **145,** 709–710.

Loken, M.R. and Herzenberg, L.A. (1975), *Ann. N.Y. Acad. Sci.*, **254,** 163.

Loken, M.R., Parks, D.R. and Herzenberg, L.A. (1977), *J. Histol. Cytol.*, **25,** 899–907.

Maartmann-Moe, K., Wang, H.S., Donald, L.J., Hamerton, J.L. and Berg, K. (1982), *Cytogenet. Cell Genet.*, **32,** 295–296.

McKenzie, I.F.C. and Potter, T. (1979), *Adv. Immunol.*, **27,** 179–328.

Mercer, W.E. and Baserga, R. (1982), in *Techniques in Somatic Cell Genetics.* (J.W. Shay, ed.), Plenum Press, New York and London.

Meruelo, D., Offer, M. and Rossomando, A. (1982), *Proc. Natl. Acad. Sci. U.S.A.*, **79,** 7460–7464.

Muller, C.R., Migl, B., Traupe, H. and Ropers, H.H. (1980), *Hum. Genet.*, **54,** 197–199.

Nabholz, M., Miggiano, V. and Bodmer, W.F. (1969), *Nature (London)*, **223,** 358–363.

Neff, J.M. and Enders, S.F. (1968), *Proc. Soc. Exp. Biol. Med.*, **127,** 260–267.

Norwood, T.H. and Zeigler, C.J. (1982), in *Techniques in Somatic Cell Genetics* (J.W. Shay, ed.), Plenum Press, New York and London.

O'Brien, S.J., Simonson, J.M. and Eichelberger, M. (1982), in *Techniques in Somatic Cell Genetics* (J.W. Shay, ed.), Plenum Press, New York and London.

Oda, M. and Puck, T.T. (1961), *J. Exp. Med.*, **113,** 599–610.

Okada, Y. (1958), *Biken's J.*, **1,** 103–110.

Okada, Y. (1962), *Exp. Cell Res.*, **26,** 98–107.

Ott, J., Schrott, H.G., Goldstein, J.L., Hazzard, W.L., Allen, F.H., Jr., Falk, C.T. and Motulsky, A.G. (1974), *Am. J. Hum. Genet.*, **26,** 598–603.

Parham, P. and Strominger, J. (eds) (1982), *Receptors and Recognition*, B14, Chapman and Hall, London.

Pearson, P.C. and Bobrow, M. (1970), *Nature (London),* **226,** 959–961.
Peters, P.G.M., Kamarck, M.E., Hemler, M.E., Strominger, J. and Ruddle, F.H. (1982), *Somat. Cell Genet.,* **8,** 825–834.
Pontecorvo, G. (1975), *Somat. Cell Genet.,* **1,** 397–400.
Puck, T.T., Wuthier, P., Jones, C. and Kao, F. (1971), *Proc. Natl. Acad. Sci. U.S.A.,* **68,** 3102–3106.
Race, R.R. and Sanger, R. (1975), *Blood Groups in Man,* 6th edn, Blackwell Scientific, Oxford and Edinburgh.
Reinherz, E.L. and Schlossman, S.F. (1980), *Cell* **19,** 821–827.
Reinherz, E., Kung, P.C., Goldstein, G., Levey, R.H. and Schlossman, S.F. (1980), *Proc. Natl. Acad. Sci. U.S.A.* **77,** 1588–1592.
Roth, R.A., Cassell, D.J., Wong, K.Y., Maddux, B.A. and Goldfine, I.D. (1982), *Proc. Natl. Acad. Sci. U.S.A.,* **79,** 7312–7316.
Rowley, J.D. (1983) *Nature (London),* **301,** 290–291.
Sakaguchi, A.Y. and Shows, T.B. (1982), *Somat. Cell Genet.,* **8,** 83–95.
Santachiara, S.A., Nabholz, M., Miggiano, V., Darlington, A.J. and Bodmer, W.F. (1970), *Nature (London),* **227,** 248–251.
Seabright, M. (1972), *Chromosoma,* **36,** 204–210.
Seravalli, E., Schwab, R., Pernis, B. and Siniscalco, M. (1978), *Cytogenet. Cell Genet.,* **22,** 260–264.
Shapiro, L.J., Mohandas, T., Weiss, R. and Romeo, G. (1979), *Science,* **204,** 1224–1226.
Shimizu, N., Behzadian, M.A. and Shimizu, Y. (1980), *Proc. Natl. Acad. Sci. U.S.A.,* **77,** 3600–3604.
Simon, M., Alexander, J-L., Fauchet, R., Genetet, B. and Bourel, M. (1980), *Prog. Med. Genet.,* **4,** 135–168.
Sixth International Workshop on Human Gene Mapping, Oslo (1981), *Birth Defects*: Orig. Article Series, 18, 2, (1982) and *Cytogenet. Cell Genet.,* **32,** (1982).
Stanners, C.P., Lam, T., Chamberlain, J.W., Stewart, S.G. and Price, G.B. (1981), *Cell,* **27,** 211–221.
Sutherland, R., Delia, D., Scheider, C., Newman, R., Kemshead, J. and Greaves, M. (1981), *Proc. Natl. Acad. Sci. U.S.A.,* **78,** 4515–4518.
Szybalski, W., Szybalska, E.H. and Ragni, G. (1962), *Natl. Cancer Inst. Monogr.,* **7,** 75–89.
Tunnacliffe, A., Goodfellow, P., Banting, G., Solomon, E., Knowles, B.B. and Andrews, P. (1983a), *Somat. Cell. Genet.,* **9,** 629–642.
Tunnacliffe, A., Jones, C. and Goodfellow, P. (1983b), *Immunology Today,* **4,** 230–233.
van Someren, H., Westerveld, A., Hagemeijer, A., Mees, J.R., Meera Kahn, P. and Zaalberg, O.P. (1974), *Proc. Natl. Acad. Sci. U.S.A.,* **71,** 962–965.
Vora, S., Durham, S., de Martinville, B., George, D.L. and Francke, U. (1982), *Somat. Cell Genet.,* **8,** 95–104.
Waterfield, M.D., Mayes, E.L.V., Stroobant, P., Bennet, P.L.P., Young, S., Goodfellow, P.N., Banting, G.S. and Ozanne, B. (1982), *J. Cell Biochem.,* **20,** 149–161.
Weiss, M.C. and Green, H. (1967), Proc. Natl. Acad. Sci. U.S.A., **58,** 1104–1111.
Whitehead, A.S., Solomon, E., Chambers, S., Bodmer, W.F., Povey, S. and Fey, G. (1982), *Proc. Natl. Acad. Sci. U.S.A.,* **79,** 5021–5025.
Williams, A.F. (1977), *Contemp. Top. Mol. Immunol.,* **6,** 83–116.

4 Molecular Genetics of the HLA Region

JOHN TROWSDALE, JANET LEE and
WALTER F. BODMER

EDITOR'S INTRODUCTION

The major histocompatibility complex (MHC), the *HLA* region in humans and *H-2* in mouse, has been the subject of such intensive study for the last decade that several books have been devoted to this subject alone (for example, *Receptors and Recognition Series B,* Volume 13). Initial interest was spurred by the role of the MHC in transplantation rejection and by the discovery of very high levels of polymorphism at several loci in this region. Genetic studies on the MHC extend from the population level to nucleotide sequences. Population studies have contributed to anthropology indicating racial and geographic origins. They have also uncovered disease associations with particular antigens and antigenic combinations. Studies at the whole-organism level have led to an understanding of the rules of tissue transplantation and to the discovery of genes in the MHC which regulate immune responses and produce complement components. Cellular studies have defined the role of these MHC products in controlling cellular interactions in the immune response both in T cell killing (class I products or classical histocompatibility antigens) and in lymphocyte–lymphocyte and lymphocyte–accessory cell interaction (class II products or Ia antigens). This regulatory role of the MHC products has prompted the suggestion that the complement components (class III products) may also have a regulatory role in cell–cell interaction in the immune response.

With the advent of the molecular genetic revolution and the possibility of studying genes directly at the nucleotide level the MHC has become a model for the detailed analysis of a complex mammalian locus. John Trowsdale, Janet Lee and Walter Bodmer describe in Chapter 4 the molecular analysis of the MHC. This analysis has implicated a new mechanism for generating class I product variants and has demonstrated that class I and class II products are evolutionarily related to each other and immunoglobulin. In addition the molecular techniques offer novel approaches to studying the function of cell surface molecules, such as MHC products, by the directed expression of manipulated genes.

Acknowledgements

We thank the following colleagues for helpful discussions and for help in preparing the manuscript: P. Travers, C. Furse, M. Carroll, R.R. Porter, D. Campbell, P. Goodfellow, R. Spielman.

Genetic Analysis of the Cell Surface
(*Receptors and Recognition,* Series B, Volume 16)
Edited by P. Goodfellow
Published in 1984 by Chapman and Hall, 11 New Fetter Lane, London EC4P 4EE

4.1 INTRODUCTION – THE *HLA* REGION

The *HLA* region, on the short arm of human chromosome 6 (Franke and Pellegrino, 1977; Goodfellow *et al.*, 1982), contains genes for three major groups of products (Fig. 4.1). These are: class I, typified by HLA-ABC antigens; class II, typified by HLA-DR antigens; and, class III, complement components. A rough estimate of the size of the region is about 2.5×10^6 base-pairs – enough DNA, making various assumptions, to code for around 100 products. These are mainly polymorphic cell surface glycoproteins associated with functions in the immune system.

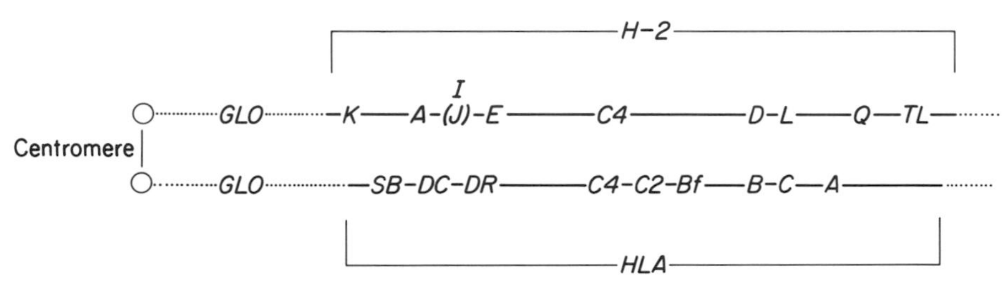

Fig. 4.1 Schematic map of the *HLA* region, on the short arm of human chromosome 6. The *H-2* region, on mouse chromosome 17, is given for comparison. The HLA class I region is represented by the three known loci, *HLA-A, B* and *C* but contains multiple, highly related, genes and pseudogenes, most of which are probably equivalent to the mouse *Q* and *TL* loci, and may, by analogy, map to the right of the *HLA-A* locus. There are at least 36 class I genes in Balb/c mice (Steinmetz *et al.*, 1982b), including 10 in the *Qa* region and 21 in *TL* (Steinmetz, personal communication), and the human system could be equally complex (Section 4.3). The class II region is represented by the three established loci, *DR, DC* and *SB*. The *SB* locus is centromeric to the others but the order of *DR* and *DC* is not established. Note that since *I-A* and *I-E* are probably the equivalent mouse genes corresponding to *DC* and *DR*, respectively, the order shown would be consistent in the two species. *DC* is one of a series of names that may refer to the same locus or to closely related genes (*DS, MB, MT, LB*) (Goyert *et al.*, 1982; Dick, 1982; Trowsdale *et al.*, 1983b), but might also be more complex. We have opted for the simplest schematic map at the moment, in view of the fact that each class II product is a dimer of products from two not necessarily contiguous genes (α and β), the precise order of which will only be established when DNA clones have been linked across the whole region. There are at least two *C4* loci, and the order of the complement component genes is not yet clear. There is about 2 to 3% recombination between the extremes of the *HLA* region corresponding to about one-thousandth of the human genome, or about 2.5×10^6 nucleotide pairs. *GL0* refers to the enzyme red cell glyoxylase.

HLA research is a rapidly moving field and this short chapter is not intended to be a comprehensive review of the literature. Rather, it introduces those aspects of *HLA* that have been illuminated by DNA cloning studies. Further background may be found in the following reviews; Alper (1981); Benacerraf (1981); Bodmer (1978, 1981); Klein (1979); Murphy (1981); Parham and Strominger (1982); Ploegh *et al.* (1981); Shackelford *et al.* (1982); Strominger *et al.* (1981). DNA-cloning studies in the mouse *H-2* and *Ia* regions have been noted only cursorily in this chapter. Good starting points for the mouse literature are recent papers by Steinmetz *et al.* (1982b), Hood *et al.* (1982), Mellor *et al.* (1982) and Nathenson *et al.* (1981).

4.2 ISOLATION OF cDNA CLONES

Although genes can be isolated directly from genomic DNA, a more practical approach to the isolation of low-copy-number genes is to obtain cDNA clones. Sequences for a single copy gene, represented, for example, by 0.1% of the total mRNA, and thus in 1/1000 cDNA clones, will be present at a frequency of around two orders of magnitude lower in a genomic DNA library. And, once a cDNA probe has been isolated it is then a relatively simple task to probe genomic DNA banks to isolate the corresponding gene sequences, by hybridization. A variety of methods has been used to obtain cDNA probes for *HLA* region products and these are listed in Table 4.1.

Table 4.1 Methods used to isolate cDNA clones for HLA products

Product		Method	Reference
Class I	HLA−B	Hybridization–selection	Ploegh *et al.* (1980)
	HLA−B	Primer extension with ddNTP synthesis	Sood *et al.* (1981)
	HLA−A2	Hybridization–selection	Bodmer *et al.* (1982)
Class II	HLA−DRα	Hybridization–selection	Lee *et al.* (1982a) Gustafsson *et al.* (1982) Wake *et al.* (1982b)
	HLA−DRα	Oligonucleotide probe used directly on to cDNA libraries	Stetler *et al.* (1982)
	HLA−DRα	Immunopurification of polysomes with monoclonal antibody	Korman *et al.* (1982)
	HLA−DRβ	Expression assay in frog oocytes	Long *et al.* (1982)
	HLA−DRβ	Hybridization–selection	Wiman *et al.* (1982)
Class III	C4	Oligonucleotide probe used directly on to cDNA libraries	Carroll and Porter (1983)
	C2	Oligonucleotide probe used directly on to cDNA libraries	Campbell and Porter (1983)
	Bf	Oligonucleotide probe used directly on to cDNA libraries	Woods *et al.* (1982)

Several groups at first opted for hybridization–selection. For example, Ploegh *et al.* (1980) isolated a cDNA clone for a class I antigen by partially purifying *HLA* mRNA 20–40-fold using agarose gel electrophoresis. The mRNA was assayed by synthesis of the protein in a cell free system. Double-stranded cDNA was then made and inserted into plasmids which were subsequently transformed into *E. coli.* The transformants were screened individually for the presence of *HLA* sequences by the ability of their DNA, immobilized on to nitrocellulose filters, to retain *HLA* mRNA specifically. This was monitored, after elution from the filters, by *in vitro* translation and immunoprecipitation. Lee *et al.* (1980, 1982a) isolated cDNA clones for the HLA-DRα chain by a similar technique, but enriched the mRNA for the required sequences by first isolating it from membrane-bound polyribosomes, and then fractionating by velocity sedimentation in sucrose gradients. DNA from the clones was then covalently attached to diazotized paper squares for the hybridization–selection procedure.

Sood *et al.* (1981) used a novel method of primer extension, using an 11-nucleotide primer coupled with dideoxynucleoside triphosphate-terminated cDNA synthesis to make a short, specific class I cDNA probe. The short oligonucleotide primer was effective because the sizes and heterogeneity of the cDNA products were limited by these procedures. Other laboratories, listed in Table 4.1, used longer oligonucleotide probes to screen cDNA libraries directly.

Other effective methods have included enrichment of specific mRNA by immunoprecipitation of polysomes using a monoclonal antibody (Korman *et al.*, 1982), and the exploitation of an expression assay in frog oocytes (Long *et al.*, 1982). The eggs were injected with a combination of positively selected mRNAs from cDNA clones for HLA−DRα and invariant (I) chains (the invariant chain is a 30000-M_r polypeptide thought to be necessary for some stage in the assembly or intracellular transport of DR antigens, that is not coded for on chromosome 6, and is not expressed on the cell surface in conjunction with DR molecules – see Section 4.3.) Correct assembly and subsequent recognition of the class II product with monoclonal antibodies was made dependent on the presence of mRNA positively selected from a β-chain clone, which was thereby detected, and isolated.

4.3 CLASS I GENES AND THEIR PRODUCTS – HLA−ABC

The HLA−ABC, or class I, antigens are dimers, made up of a 43000-M_r glycoprotein, which spans the cell membrane, non-covalently associated with the 12000-M_r protein β_2-microglobulin, encoded on chromosome 15 (Goodfellow *et al.,* 1975). There are multiple class I gene sequences in the *HLA* region. However, the originally described and major expressed loci, whose products are detected serologically, comprise three allelic series: *HLA−A*, *−B* and *−C*. The HLA−ABC antigens have a broad tissue distribution and are present on most nucleated cells, and not on red blood cells. They present a major barrier to tissue transplantation between individuals and are highly polymorphic, (see reviews cited in Section 4.1 for details). Other known human class I antigens are probably equivalent to the mouse Qa and TL products, and are generally associated with thymocytes (Cotner *et al.*, 1981; Ziegler and Milstein, 1979; Knowles and Bodmer, 1982).

Studies of the *HLA−ABC* and *H−2KDL* class I protein products by Strominger and Nathenson and their colleagues have shown that in each of these molecules the 43 000-M_r polypeptide can be divided into three external domains (Fig. 4.2). The $\alpha 1$ domain consists of 90 residues and carries an attachment site for an asparagine-linked carbohydrate side chain at position 86, at the canonical sequence Asn-X-Thr/Ser. The $\alpha 2$ and $\alpha 3$ domains both contain disulphide loops. The amino acid sequence of the $\alpha 3$ domain indicates a significant level of homology with Ig constant regions, and with β_2-microglobulin (Orr *et al.*, 1979; Tragardh *et al.*, 1979, 1980; Peterson *et al.*, 1972). The hydrophobic and uncharged amino acids that comprise the membrane-spanning portion of class I molecules are remarkably conserved between genes and species, while the cytoplasmic portions show considerable differences, particularly between species. From the comparisons of protein sequences made so far, the alloantigenic portions of the *HLA−ABC* and *H−2KDL* molecules appear to reside primarily in discrete areas of the $\alpha 1$ and $\alpha 2$ domains (Ploegh *et al.*, 1981). As will be described below, the gene structure of the class I antigens helps to confirm and extend what is already known from these protein studies.

Genomic DNA libraries have been screened, using cDNA clones as probes, to isolate clones containing class I *HLA* genes. To date, only one such human gene has been completely sequenced, from a DNA library in bacteriophage λ charon 4A vector (Malissen *et al.*, 1982a). Like most other eukaryotic genes, those for *HLA* are divided into coding sequences (exons) and non-coding sequences, known as introns. The gene described by Malissen and her colleagues contains seven exons. The first four exons coded for a signal sequence peptide, and the first, second, and third extracellular domains of the molecule ($\alpha 1$, $\alpha 2$ and $\alpha 3$ respectively) as shown in Fig. 4.2. The exon encoding the transmembrane region also includes a peptide that bridges the $\alpha 3$ and transmembrane regions (the connecting peptide) as well as the first few amino acids of the cytoplasmic region. Surprisingly, the rest of the cytoplasmic segment is also split amongst a further two exons. In some cases, e.g. *HLA−A2*, there are possibly three such exons (unpublished results), as for some mouse *H-2* genes (Steinmetz *et al.*, 1981a,b). Alternative RNA-splicing sites in this region might conceivably result in the synthesis of antigens with identical extracellular domains and different cytoplasmic domains.

There is overwhelming evidence now from both human and mouse systems that class I antigen genes have the general structure shown in Fig. 4.2, although the gene studied by Malissen *et al.* (1982a) may be a pseudogene, since the second cysteine, usually involved in one of the disulphide bridges, is replaced by a phenylalanine residue. The finding of a complete *HLA* gene in genomic DNA, contiguous with the mRNA and protein sequences, puts paid to some of the more fanciful suggestions for generation of polymorphism in the *HLA* genes. For example, it seems unlikely that much polymorphism of class I antigens could be generated by DNA rearrangements or differential splicing

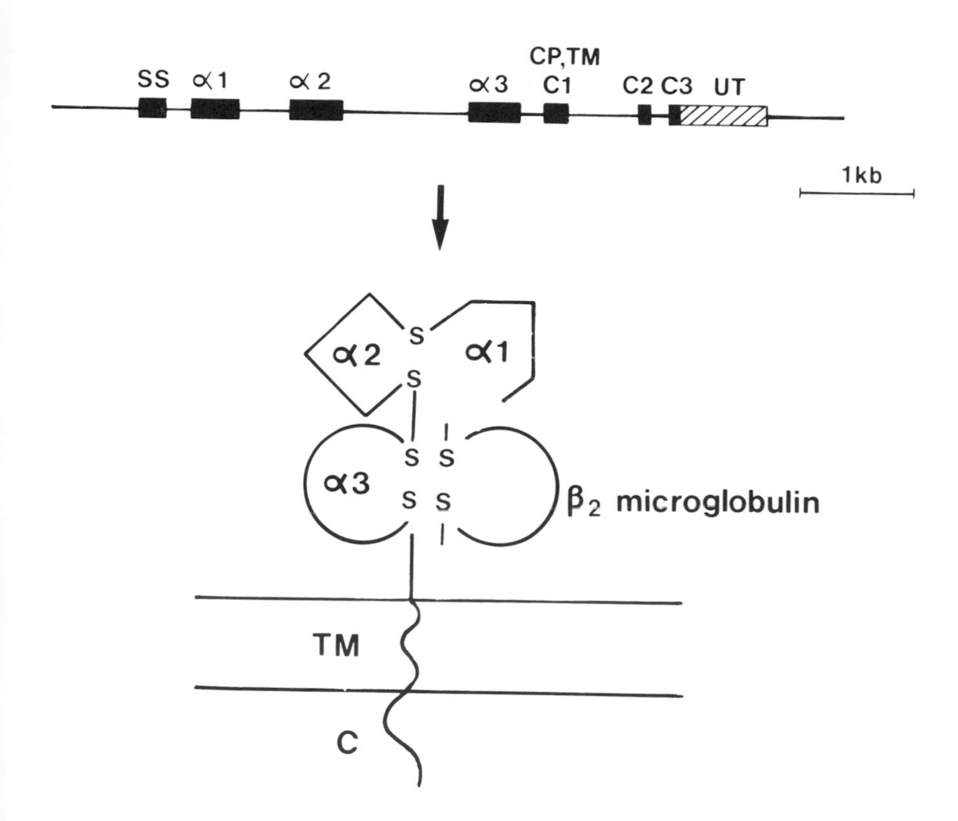

Fig. 4.2 Intron/exon organization of a class I gene and correlation with the structure of a class I transplantation antigen. The exons are indicated by filled boxes, starting with the signal sequence (SS) exon at the end of the gene which gives the 5′ end of the mRNA. CP, connecting peptide; C(1,2,3) cytoplasmic portions; TM, transmembrane region; UT, the 3′ untranslated region.

involving the use of alternative adjacent exons (Bodmer, 1981). However, it has not yet been ruled out that some polymorphisms could be associated with differences between individuals as to which particular genes of the region are expressed, or are functional, as suggested by Bodmer (1972).

Malissen *et al.* (1982b) have also isolated cosmid clones containing other human *HLA−ABC* genes. As soon as cDNA clones became available Southern blotting experiments indicated that there were a large number of highly homologous class I genes (Orr *et al.*, 1982). This was confirmed by direct cloning in mouse (Steinmetz *et al.*, 1982b) and man (Malissen *et al.*, 1982b). In contrast to the mouse, most of the ~ 40 kb human cosmid clones only contained one gene, and so it has been argued that the human genes are more widely spaced than those in the mouse. Steinmetz *et al.* (1982b) have reported 36 mouse class I genes in 837 kb of DNA; equivalent to about one gene for every 20 kb. Where there was more than one gene on one DNA fragment in the human cosmids the orientation of both genes was the same (Malissen *et al.*, 1982b). Similarly, in Balb/c mice, seven clustered genes were all oriented in the same 5′ and 3′ direction (Steinmetz *et al.*, 1982b). Though the organization of the mouse *H-2* region may well be different to that of man, it is worth noting that most of the 36 class I genes described by Steinmetz *et al.* (1981b) mapped to the *Qa* and *TL* regions; while the *K, D* and *L* loci (equivalent to *HLA−A, −B* and *−C*) probably have fewer than two genes at each locus.

Clearcut identification of the cloned gene sequences presents a problem, except when direct protein sequence data are available. Gene-transfer experiments can, however, help to identify the clones, through their products, and this technique is being used in several laboratories (in mouse, see for example, Goodenow *et al.*, 1983; Mellor *et al.*, 1982). Another way of identifying particular alleles is to search for restriction enzyme polymorphisms associated with individual alleles. In this way, Cann *et al.* (1983) found an EcoRV fragment of 8.6 kb that correlated essentially completely with the serologically defined specificity *HLA−B8*.

Although not on chromosome 6, β_2-microglobulin should be mentioned because of its importance in the structure of the class I transplantation antigens. In man, cDNA clones (Suggs *et al.*, 1981) and in mouse, both cDNA and genomic DNA clones (Parnes *et al.*, 1981; Parnes and Seidman, 1982), have been isolated. The mouse β_2-microglobulin gene contains three introns, and the bulk of the protein-coding sequence is confined to a single exon (Parnes and Seidman, 1982). Recently a single allelic β_2m variant has been described in the mouse by Michaelson *et al.* (1980), though so far no β_2m polymorphism has been observed in humans. In the mouse a series of polymorphisms for cell surface antigens has been shown to be closely linked to the β_2m locus (Meruelo *et al.*, 1982). It will be interesting to see whether human clones, when used as probes, can identify polymorphisms which are the counterparts of these mouse determinants.

4.4 CLASS II GENES AND THEIR PRODUCTS – THE *HLA–D* REGION

The HLA–D related, or class II, antigens are oligomeric cell surface glycoprotein molecules composed of non-covalently associated $M_r \sim 33000$ and ~ 28000 subunits, the α and β chains respectively. Both α and β chains span the cell membrane (Walsh and Crumpton, 1977). The genes for both chains are in the *HLA–D* region of chromosome 6 in a cluster of related sequences (Fig. 4.1). The known loci include at least three distinct allelic series: *HLA–DR*, *–DC* (also known by a number of other names which probably refer to the same locus or to related loci – see legend to Fig. 4.1, and Dick, 1982; Guy and Van Heyningen, 1982; Trowsdale *et al.*, 1983a,b) and *SB* (Shaw *et al.*, 1981; Kavathas *et al.*, 1981). In each case the major electrophoretic polymorphism seems to reside in the β chains (Charron and McDevitt, 1979; Shackelford and Strominger, 1980; de Kretser *et al.*, 1982a,b). The class II antigens were originally described on B (and not T) lymphocytes and monocytes or macrophages, but are now known to be associated with certain other tissues including in particular activated lymphocytes (reviewed recently for the mouse Ia system by Nixon *et al.*, 1982).

The first probes in the *HLA–D* region to be isolated were cDNA clones for the HLA–DRα chain, homologous to the α chain of the mouse *I–E* product (see Table 4.1). The data from the *HLA–DRα* cDNA clones (Section 4.2) and derived genomic clones provided the first complete protein sequence, of which there was previously only fragmentary knowledge (Lee *et al.*, 1982a; Korman *et al.*, 1982; Larhammar *et al.*, 1982b). The schematic structure of the *HLA-DRα* chain gene and its protein product are shown in Fig. 4.3.

The gene consists of five exons, the first encoding the signal sequence and a separate exon for the major part of each of two external protein domains, $\alpha 1$ and $\alpha 2$, the latter (membrane proximal) containing a disulphide bond. The next exon encodes a peptide connecting the $\alpha 2$ domain to the transmembrane segment, and the C-terminal cytoplasmic tail, while all but the first 12 nucleotides of the 3′ untranslated region are contained in a further, separate exon. There are two polyadenylation signals (AATAAA) in the last exon, separated by ~100 bp.

The proposed HLA-DRα chain molecule contains 229 amino acids. The predicted $\alpha 1$ domain contains 82 amino acids, including one glycosylation site, Asp-X-Thr/Ser. As mentioned above, the predicted $\alpha 2$ domain contains two cysteine residues for disulphide bonding, in a domain of 94 amino acids, as well as a further glycosylation site. It also shows significant homology to Ig constant region domains as do β_2m and the $\alpha 3$ domain of HLA-ABC, these together comprising a family of structurally and evolutionarily related domains (see Section 4.7 and Fig. 4.4). Twenty-three hydrophobic amino acids comprise the

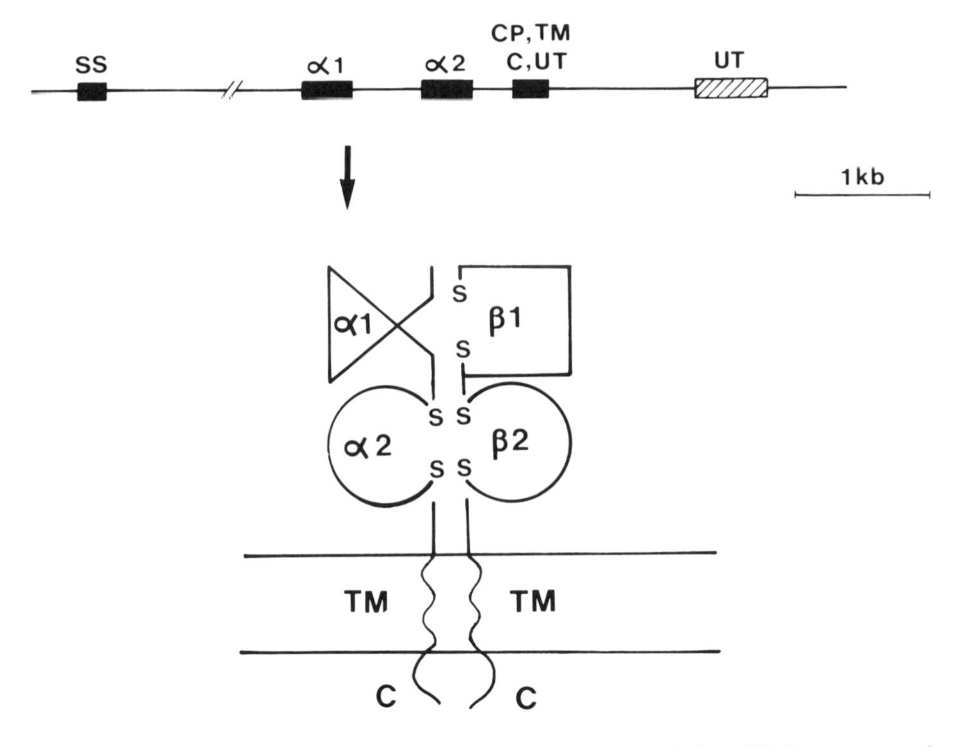

Fig. 4.3 Intron/exon organization of a class II α gene and correlation with the structure of a class II antigen. The exons are indicated by filled boxes, starting with the signal sequence (SS) exon. CP, connecting peptide; TM, transmembrane region.

membrane-spanning portion of the molecule, including a cysteine residue. The cytoplasmic tail, containing only 15 amino acids, is slightly shorter than the equivalent region in class I molecules. Korman *et al.* (1982) pointed out the fact that the stop codon for the HLA-DRα chain is on a different exon to the 3′ untranslated region.

cDNA and genomic DNA clones have also been isolated for other *HLA-D* region α chain sequences, by hybridization to *HLA-DR*α chain probes under relatively non-stringent conditions (Trowsdale *et al.*, 1983a; Auffray *et al.*, 1982). One of the cDNA clones (Auffray *et al.*, 1982) and a cosmid genomic DNA clone (Trowsdale *et al.*, 1983a) correspond to a *DC* locus α chain. From sequence analysis, Auffray *et al.* (1982) ascribed their cDNA clone to the HLA-DC1 specificity (Bono and Strominger, 1982). The clones indicate a sequence homology with mouse I–A α chains and are probably homologous in the two different species (Bono and Strominger, 1982; Goyert *et al.*, 1982). Comparison of the *HLA-DR* and *HLA-DC*α chain sequences indicates similar structures which, however, only share 57% overall homology (Auffray *et al.*,

```
DR-α2      ITNVPPEVTVLTNSPVELREPNVLICFIDKFTPPVVNVTWLRNGKPVTTGVSETVFLPREDHLFRKFHYLPFLPSTEDVYDCRVEHWGLDEPLLKHWEFDA
DC-α2       NEVPEVTVFSKSPVTLGNPNTLICLVDNIFPPVVNITWLSNGHSVTGGVSETSFLSKSDHSFFKISYLTFLPSADEIYDCKVEHWGLDEPLLKHWEPEI
DR β2       RVQPKVTVYPSKTQPLQHHNLLVCSVSGFYPGSIEVRWFLNGEEKAGGVS  TGLIQDDWTFQTLVMLETVPRSGEVYTCQVEHPSVTSPLTVEWRARS
DC β2       RVEPTVTISPSRTEALNHHNLLVCSVTDFYPAQIKVRWFRNDEETAGGVS  TPLIRNDWTFQILVMLEMTPQRGDVYTCHVEHPSLQSPITVEWRAQS
HLA B7α3    RADPPKTHVTHHPIS DHEATLRCWALGFYPAEITLTWQRDGEDQTQDTELVETRPAGDRTFQKWAAVVVPSGEGQRYTCHVQHEGLPKPLTLGWEPSS
ΨHLA α3     RADPPKTHMTHHPIS DHEATLRCWALGFYPAEITLTWQRDGEDQTQDTELVETRPAGDGTFQKWAAVVVPSGEEQRYTCHVQHEGLPEPLTLRWEPSS
β2m         QRTPKIQVYSRHPAENGKSNFLNCYVSGFHPSDIEVDLLKNGERIEKVEHS DLSFSKDWSFYLLYYTEFTPTEKDEYHCRVNHVTLSQPKIVKWDRDM
Cλ         QPKAAPSVTLFPPSSEELQNKATLVCLISDFYPGAVTVAWKADSSPVKAGVETTTPSKQSNNKYAASSYLSLTPEQHKSYSCQVTHEGSTVEKTVAPTECS
           L110     L120     A L130    L140      L150      L160      L170      L180    WKS L190    L200     L210
```

(Arrows above the alignment mark shared residues P, L, C, P, C, V, H; variant residues Q and G are printed above DR β2 and DC β2 at positions ~L154 and ~L169.)

Fig. 4.4 Comparison of the amino acid sequences of the membrane-proximal domains of class I and class II molecules as deduced from nucleic acid sequences with β_2-microglobulin and immunoglobulin constant regions. The sequences were from the following sources: *HLA-Drα* chain (Lee *et al.*, (1982b) *DClα* chain (Auffray *et al.*, 1982) *pDR-β-1β* chain Larhammar *et al.*, 1981, 1982b); *HLA-B7*. (Weissman, personal communication); *pseudo HLA* (Malissen *et al.*, 1982a). Amino acid residues shared by all the sequences are indicated by arrows. Residues shared by only the *HLA* region products are indicated by a black dot. Data compiled by P. Travers (personal communication).

1982). The homologies are not evenly distributed amongst the domains. The DCα2 (membrane proximal) and membrane-spanning domains were, respectively, 65% and 74% homologous to corresponding domains in the DRα chain, but the α1 domains, and the intracytoplasmic portions were much more divergent (47.1% and 47.6% respectively). Sequencing of the exon corresponding to the α2 domain in an *HLA-DC* gene indicates similar splicing donor points in both *DRα* and *DCα* genes and an expected lack of homology in intron sequences (Trowsdale *et al.*, 1983a).

A third class II α chain gene has been isolated by Trowsdale *et al.* (1983a) which could, it was proposed, correspond to the SB α chain (see Hurley *et al.*, 1982), a novel class II α chain gene, or to a pseudogene. The α2 region of this gene is also clearly related to the Ig constant region domain (Trowsdale *et al.*, 1983a). The gene is expressed in mRNA as indicated by Northern blots (Lee and Trowsdale, unpublished), and corresponding cDNA clones have been isolated (H. Ehrlich and C. Auffray, personal communications). It was shown using human–mouse somatic cell hybrids that all of the above-mentioned sequences are present on human chromosome 6 and so, presumably, in the *HLA* region (Trowsdale *et al.*, 1983a).

Thus, there is so far direct sequence evidence for three related human class II α chain genes. Blotting studies and analysis of cosmid clones, however, indicate that there are still further sequences related to class II α chain sequences in the human genome. One of these is closely related to *DCα*, observed by both Auffray *et al.* (1983) and Trowsdale *et al.* (1983a). This was demonstrated both by direct cloning and by the fact that several non-overlapping DNA probes from the *DCα* chain gene hybridized with two bands on Southern blots of human DNA and was further supported by family studies of restriction fragment polymorphisms identified on Southern blots using this clone (Trowsdale *et al.*, 1983a). In addition, other cosmid clones have been isolated which between them contain at least two further α chain genes (J. Trowsdale and J. Lee, unpublished). There could therefore be six or more such sequences in the human genome, some of which may of course be non-expressed pseudogenes. Analysis of class II proteins by two-dimensional electrophoresis calls for at least three oligomeric products (see e.g. de Kretzer *et al.*, 1982a,b), whereas, in the mouse, so far, only two class II α chain genes have been described (Steinmetz *et al.*, 1982a).

Sequences of related β chain cDNA clones have been obtained (Long *et al.*, 1983b; Larhammar *et al.*, 1982b). These sequences indicate that there are two disulphide-bonded protein domains in the β chain, and that the membrane proximal-domain is again reminiscent of an Ig constant domain (Section 4.7). There is a canonical carbohydrate attachment site in the first domain. The membrane-spanning region consists of 21 hydrophobic or uncharged amino acids, and there is a cytoplasmic tail of 10 residues. On the basis of preliminary data the structure of the *HLA-DRβ* chain genes show similarity to the α chain genes at the genomic level except that the intracytoplasmic portion is divided into at least two exons, in a manner analogous to the organization of this portion of the class I product (P. Peterson *et al.*, personal communication).

Though two-dimensional gel electrophoresis has so far clearly identified only three sets of β chains, protein-sequencing data of *HLA-D* region products isolated from homozygous cell lines has suggested that there might be at least seven related β chain sequences (Kratzin *et al.*, 1981). Blotting experiments and cDNA cloning have subsequently helped to confirm that there are a considerable number of β chain genes and more than three loci, though how many are expressed is not yet clear (Bohme *et al.*, 1983; Wake *et al.*, 1982a). The finding of multiple β-chain genes is in general agreement with data from the mouse where six β chain genes have so far been located (Steinmetz *et al.*, 1982a; and personal communication). Direct comparisons of the sequences of DRβ chain cDNA clones isolated so far also indicate that there are at least three DRβ genes. Most of the differences between DRβ chains found by Long *et al.* (1983a), but not proven to be allelic, were single amino acid substitutions.

The Southern blotting studies using the class II probes have also suggested: (1) that the α chain genes in general had relatively little restriction-enzyme polymorphism, compatible with the fact that they are more or less invariant on two-dimensional gels (de Kretzer *et al.*, 1982a,b); (2) the β chain genes showed considerable polymorphism when used in a similar way as probes (Bohme *et al.*, 1983; Wake *et al.*, 1982a). One notable exception to this pattern is that the *DCα* chain probes revealed considerable variation at the DNA level in the vicinity of the *DCα* chain gene. It is not yet established whether this corresponds to differences in the primary structure of the proteins or relates to sequences flanking the coding regions of the gene (Trowsdale *et al.*, 1983a; Auffray *et al.*, 1983). These studies also indicate how DNA probes can be effectively used to detect differences that can be exploited for genetic studies involving, for example, HLA and disease associations. Trowsdale *et al.* (1983a) and Auffray *et al.* (1983) have shown that restriction patterns detected by *DCα* probes were characteristic of different DR serological phenotypes and could be followed in a family study, while in one case at least, more than one pattern could be found with a given DR type.

The I chain gene has also been cloned recently (Long *et al.*, 1983a). It does not map to chromosome 6, in agreement with studies in the mouse which have shown that the corresponding mouse gene is not in the *H-2* region (Day and Jones, 1983). This chain is thought to be involved with intracellular processing of class II antigens and does not appear at the cell surface in association with *HLA-D* region serologically identified products. There is evidence for two I chain proteins in man, both invariant in B cells, with molecular weights of about 32 K and 35 K, but one is possibly a precursor of the other (Shackelford *et al.*, 1982; D. Charron and C. Rudd, personal communications). One report suggests that mouse invariant chains can be detected on the surface of lymphocytes using a specific monoclonal antibody (Koch *et al.*, 1982).

4.5 CLASS III GENES AND THEIR PRODUCTS–C2, C4, Bf

The other major components of the *HLA* region are the complement components: C2, C4 and factor B (Porter, 1979; Shreffler, 1982). C2, of the classical complement pathway, and factor B, of the alternative pathway, have similar, apparently homologous, structures consisting of a single polypeptide chain, $M_r \sim 95000$. The fourth component of the classical complement pathway, C4, is a protein of $M_r \sim 200000$, and is composed of three polypeptide chains: $\alpha(M_r \sim 95000)$, $\beta(M_r \sim 75000)$ and $\gamma(M_r \sim 30000)$. It is synthesized as a single polypeptide chain that undergoes proteolytic cleavage on secretion to give, sequentially, the β, α and γ chains. Population and family studies suggest that there are two separate multiallelic *C4* loci (Raum *et al.*, 1980).

Recently, a cDNA clone containing some coding sequence for the complement component C4 was identified using an oligonucleotide probe (Carroll and Porter, 1983). The cDNA clone was then used as a hybridization probe on a human cosmid library from which several cosmids containing *C4* genes were isolated. One of the cosmid clones has been studied in detail and the sequence in one exon is identical to part of the cDNA clone sequence. Further cloning studies indicate that there are at least two *C4* genes (Carroll, personal communication).

DNA clones for factor B have also been isolated using a similar approach (Campbell and Porter, 1983; Woods *et al.*, 1982). The gene encoding factor B is approximately 5 kb, while the mRNA is 2.5 kb and the coding sequence contains 11 exons. It is particularly interesting that the amino acids constituting the active site for C5 convertase are encoded in three different exons, as are the residues important for the primary and secondary binding sites (Campbell and Porter, personal communication).

4.6 OTHER GENES IN THE *HLA* REGION

The recessive inborn error of metabolism, adrenal hyperplasia, associated with a deficiency of the steroid hydroxylating enzyme 21-hydroxylase is very closely linked to *HLA* and according to some data maps in between the *HLA–B* and *–D* loci (Dupont *et al.*, 1980). However, the nature of the biochemical control from the *HLA* region is not clear and recent work (Awdeh *et al.*, 1983) suggests that the 21-hydroxylase deficiency may be associated with genetic rearrangements in or around the *HLA* region. This is a situation with similarities to the *T/t* locus mutants in the mouse (Artzt *et al.*, 1982 a,b; Fleischnick *et al.*, 1983).

Recently also, a neuraminidase activity has been mapped to the *H-2* region, but once again the genetic and molecular basis for this is not clear (Womak *et al.*, 1981). So far, there is little evidence at the molecular level for any additional novel genes in the *HLA* region. Steinmetz *et al.* (1982a) have searched for further mouse *Ia* region genes by saturating the region with overlapping cosmid clones but have been unable to assign a gene to *I-J*, a region with suggested involvement in suppressor T-cell activity (Steinmetz *et al.*, 1982a; Klein and Nagy, 1982). The controversy in relating the classic, functional *Ir* map with the cloned genes may be resolved eventually by mapping all of the active DNA sequences by overlapping cosmids. This should also reveal any gene products that might be associated with 21-hydroxylase deficiency or neuraminidase activity.

There are a few isolated reports of additional partially characterized gene products in the *H-2* region. Hiramatsu *et al.* (1982) reported an *Ia* region product in the mouse that was detected by a monoclonal antibody to T cells. Monaco and McDevitt (1982) described a large set of small polypeptides that were encoded in the *H-2* region, and there is evidence for a M_r 22 000 product, also encoded in the region (McKean, 1983). A ~600 base-pair sequence that hybridizes to a poly(A)$^+$ mRNA has been found on one of the cosmids containing *HLA*–*D* region α and β genes (probably *SB*) close to the 5′ end, or possibly in the first intron of a β gene (Trowsdale *et al.*, unpublished). It is not yet clear what the functional significance of this sequence is.

Many of the diseases shown to be associated with the *HLA* region, such as ankylosing spondylitis, rheumatoid arthritis, juvenile insulin-dependent diabetes, myasthenia gravis, and systemic lupus erythematosis, have a clear immune-related aetiology. These associations can probably be accounted for by differences among the known types of products discussed so far, in terms of immunological mechanisms associated with autoimmunity. R.R. Porter (personal communication) has pointed out that some of the disease associations with the *HLA* region could also be due to over- or under-effective clearance of antigen–antibody complexes, connected with differences in activity of complement components, perhaps particularly C4. In addition, C2 deficiency has been associated with certain forms of systemic lupus erythematosis.

Thus estimates of the number of genes in the *HLA* region by recombination, and the analysis of genomic DNA clones in both mouse (Steinmetz *et al.*, 1982a) and man (Malissen *et al.*, 1982b; Trowsdale *et al.*, 1983a,b), as described in the previous section, lead to a current concept of the region as containing around 30 class I gene sequences, at least six class II α genes, and perhaps more β chain genes, plus at least four large complement genes. Those cosmid clones that have been analysed containing most of these genes have not as yet been shown to contain any non-*HLA* class I, II or III genes and it seems likely that few are to be found in this region. The next few years should see a precise map for the whole region and sequences will be associated with known products and functions. DNA probes for the *HLA* region, which are being used to attempt to identify the genes, or gene regions (Cann *et al.*, 1983), accounting for some of the *HLA*-associated diseases should thus help to localize precisely at the DNA level the genes involved in the *HLA*-associated disease mechanisms, and, as argued above, many of these may be in or involve expression of known products.

4.7 EVOLUTIONARY RELATIONSHIPS AND RECOMBINATION

Structurally, the class I HLA antigens are similar to each other, regardless of the allele, and regardless of the locus from which they are derived. As mentioned earlier, the amino acid variability that confers alloantigenicity occurs mainly in discrete regions of the molecules that, in most cases, presumably do not influence their overall conformation. *HLA−A2* and *−B8*, for example, were shown by direct sequencing to be about 90% related (Ploegh *et al.*, 1981). The mouse cloning studies indicate a similar situation. Domain-specific probes used in hybridization studies have shown that the $\alpha3$ domain exon is highly conserved amongst the different class I genes while the other two extracellular domains show more divergence (Steinmetz *et al.*, 1981a). There are indications, however, that the human equivalent of the mouse TL product is significantly different from the human HLA−ABC product in size, and degree of glycosylation (Cotner *et al.*, 1981; Knowles and Bodmer, 1982; R. Knowles, personal communication).

In general, this is also true of the class II antigens. Early experiments showed that class II α chain genes were more divergent than their class I counterparts, since, under stringent conditions, *HLA−DR*α cDNA probes only hybridized to a single DNA species (Lee *et al.*, 1982a). Further studies have shown that the DCα and DRα chains have diverged considerably and now show an overall similarity of only about 50%. At least some of the class II β chains studied so far, however, appear to be more homologous, since a probe made from a single β chain sequence detects those for several other genes on Southern blots of human genomic DNA under stringent conditions (Wake *et al.*, 1982a; Bohme *et al.*, 1983).

In all of these comparisons the most striking relationships are those between the membrane-proximal domains of all the class I and class II molecules so far described ($\alpha3$ of class I, and $\alpha2$ and $\beta2$ domains of class II) with each other and with β_2-microglobulin and immunoglobulin constant domains (Fig. 4.4). Different constant-region domains have about as much homology to each other by amino acid sequence as they do to the HLA antigens, as indicated in Fig. 4.4. The closest homology is found around the two cysteines, presumably because structural constraints require the maintenance of the antibody-like folds in order to permit these domains to interact with each other, much as the heavy- and light-chain domains interact in immunoglobulins.

All of the data so far are consistent with divergent evolution of the above-mentioned cell surface products from a common ancestral sequence. It is tempting to speculate that there is still a strong functional link between the *HLA* region products and immunoglobulins (Bodmer, 1972; Jensenius and Williams, 1982; Burnett, 1970; Gally and Edelman, 1972), though possibly via such functions as complement binding or even antigen recognition (Lee *et al.*, 1982b).

Homology has also been found between the $\alpha1$ and $\alpha2$ domains of class I molecules (Tragardh *et al.*, 1980). The $\beta1$ domain of a class II product also shows a weak relationship to class I $\alpha1$ domains and an even weaker one with the $\alpha2$ domain (Larhammar *et al.*, 1982b; Travers, personal communication). Thus the class I $\alpha1$ and $\alpha2$ and class II $\beta2$ domains might form another, loosely related, family of domains. The molecular relationships between the class I and II HLA molecules described above signify a remarkable pattern of evolution of a number of the major cell surface products through duplication of a small number of primordial sequences.

As yet there is no apparent homology, at the sequence level, between the class III genes and the other structures in the *HLA* region. Amino acid sequence data for C2 and factor B, confirmed by analysis of their corresponding genes, have indicated strong similarity between the two molecules, particularly in the regions of the primary binding sites (Campbell and Porter, personal communication) consistent with their expected origin by duplication from a common ancestor. There is thought to be some relationship between complement components C3 and C4 (Porter, 1979). But whereas C4 is coded for in the *HLA* region, the *C3* gene is on chromosome 19 (Whitehead *et al.*, 1982). This suggests that the complement genes may have been brought into the *HLA* region from elsewhere, and then would not be part of the same evolutionarily related set of products as the class I and class II products. The linkage between the three sets of products may, nevertheless, have been selected for because of advantageous functional interactions associated with having such a set of genes in the same cluster.

A puzzling observation, which was much discussed before the gene cloning work, was the difficulty in distinguishing biochemically whether an antigen was from the *A* or the *B* locus. A striking illustration of this problem was the behaviour of the W6/32 antibody, which bound to class I antigens from both *A* and *B* (and other) loci, but to neither of the analogous H-2 antigens, from the *K* and *D* loci (see Bodmer, 1981). It seemed unlikely that duplication of a primordial class I gene had taken place independently in both species. One explanation of the paradox was that class I genes shared exons, although this speculation was hard to reconcile with the genetic map. The answer now appears much more likely to be to invoke gene conversion, in the most general sense, or double unequal crossing-over, as a mechanism for maintaining some structural similarity amongst the products of different loci. These mechanisms can exchange or transpose sequences within a region and can lead to generation of complex differences at a stroke, without changing order, and so may lead to similar sequences being found possibly in several genes in the same cluster (see, for example, Baltimore, 1981; Bodmer, 1981 and Dover 1982). Recent data at the protein (Nathenson *et al.*, 1981; Pease *et al.*, 1983) and DNA level (Weiss *et al.*, 1983a,b) have supported these ideas by showing that a number of H-2 mutants seem to involve displacement of small sequences from one *H-2* gene to another.

Goodenow *et al.* (1983), in addition, have found that incomplete transplantation antigens were expressed in TK mouse L cells after transformation with high concentrations of DNA, suggesting that novel, comparatively frequent, recombination mechanisms may apply to class I genes, since other genes do not normally home to specific chromosomes when introduced into cells by transfection (Cohen and Murphey-Corb, 1983). However, the success in selecting for homologous recombination in the *H-2* genes may simply be due to the use of very high DNA concentrations.

It is clear that there are a variety of complex mechanisms at the DNA level by which new variations can be produced and on which natural selection may act to maintain polymorphism and promote evolutionary change. These mechanisms may explain puzzling phenomena, such as the linkage disequilibrium that may occur between related serological determinants in nearby genes, as in the case of the *HLA–DR* and *DC* allelic differences. Such mechanisms, including in particular asymmetrical gene conversion, may also account for substitution of complex genetic differences in evolution (Bodmer, 1981) by a set of processes which have been called molecular drive (see Dover, 1982). The basic driving force, however, behind the establishment and maintenance of the extraordinarily high level of polymorphism of some of the HLA and H-2 products probably still involves selective mechanisms, along the lines discussed by Bodmer (1972) and others. It is possible that if this selection operates mainly via immune response differences, to protect the human species against attack by novel pathogens, mechanisms for generating novel variants that can give rise to the appropriate immune response differences at a sufficiently high rate may be selected for, as a secondary evolutionary phenomenon. Thus, there may be pressure to increase the number of genes in the *HLA* region so as to increase the rate of variant production, in order to allow more flexible response to novel pathogens. This follows a principle analogous to that of the antigenic variation described for protozoan parasites such as the trypanozomes (see, for example, Borst *et al.*, 1981). Since these complex mechanisms for generating variants can give rise to a number of differences in sequence, only one of which may be functionally relevant, the polymorphic regions of the gene (which are those of greatest functional significance for immune response to, for example, pathogens) will tend to be variable at a number of sites, as well as those directly selected for. Undoubtedly, the complexity of the region and the variety of processes by which variants can be produced, together with the selective importance of immune response differences, gives rise to an extraordinary variety of evolutionary developments of the region, both within and between species.

4.8 CONCLUSION

In this brief survey we have illustrated how, in the space of three years, gene cloning has revolutionized studies of *HLA*. This new technology has provided the structures and detailed sequence for genes of all three classes. It has confirmed and in some cases revealed the remarkable evolutionary relationships between all of the class I and II genes with β_2-microglobulin and immunoglobulins. Along with direct protein sequencing, it has uncovered recombination mechanisms, i.e. gene conversion, that were not until recently given serious consideration in human genetics. Current studies are directed at exploiting these techniques in the directed expression of HLA products, to study their function and relationships both to disease and to cellular interaction of surface receptors.

REFERENCES

Alper, C.A. (1981), in *The Role of the Major Histocompatibility Complex in Immunobiology* (M.E. Dorf, ed.), Garland SPTM, New York, pp. 173–220.

Artzt, K., McCorkick, P. and Bennett, D. (1982a), *Cell,* **28,** 463–470.

Artzt, K., Shim, H. and Bennett, D. (1982b), *Cell,* **28,** 471–476.

Auffray, C., Korman, A.J., Roux-Dosseto, M., Bono, R. and Strominger, J. (1982), *Proc. Natl. Acad. Sci. U.S.A.,* **79,** 6337–6341.

Auffray, C., Ben-Nun, A., Roux-Dosseto, M., Germain, R.N., Seidman, J.G. and Strominger, J. (1983), *EMBO Journal,* **2,** 121–124.

Awdeh, Z.L., Raum, D., Yunis, E.J. and Alper, C.A. (1983), *Proc. Natl. Acad. Sci. U.S.A.,* **80,** 259–263.

Baltimore, D. (1981), *Cell,* **24,** 592–594.

Benacerraf, B. (1981), in *The Role of the Major Histocompatibility Complex in Immunobiology* (M.E. Dorf, ed.) Garland SPTM, New York, pp. 255–259.

Bodmer, W.F. (1972), *Nature (London),* **237,** 139–145.

Bodmer, W.F. (1978), *HLA–A super supergene. Harvey Lectures 1976–1977,* Academic Press, New York, pp. 91–138.

Bodmer, W.F. (1981), in *Mammalian Genetics and Cancer: The Jackson Laboratory, Fiftieth Anniversary Symposium,* Alan R. Liss, New York, pp. 213–239.

Bodmer, W.F., Carey, J., Jenkins, J., Lee, J. and Trowsdale, J. (1982), in *Primary and Tertiary Structure of Nucleic Acids and Cancer Research* (M. Wiwa, S. Nishimura, A. Rich, D.G. Söll and T. Sugimura, eds), Japan Science Society Press, Tokyo, pp. 307–320.

Bohme, J., Owerbach, D., Denaro, M., Lernmark, A., Peterson, P.A. and Rask, L. (1983), *Nature (London)*, **301**, 82–84.

Bono, R. and Strominger, J.L. (1982), *Nature (London)*, **299**, 836–838.

Borst, P., Frasch, A.C.C., Bernards, A., Van der Ploeg, L.H.T., Hoeijmakers, J.H.J., Arnberg, A.C. and Cross, G.A.M. (1981), *Cold Spring Harbor Symp. Quant. Biol.*, **45**, 935–943.

Burnett, F.M. (1970), *Nature (London)*, **226**, 123–126.

Campbell, D. and Porter, R.R. (1983), *Proc. Natl. Acad. Sci. U.S.A.*, (in press).

Cann, H.M., Ascanio, L., Paul, P., Marcadet, A. and Dausset, J. (1983), *Proc. Natl. Acad. Sci. U.S.A.*, **80**, 1665–1668.

Carroll, M.C. and Porter, R.R. (1983), *Proc. Natl. Acad. Sci. U.S.A.*, **80**, 264–267.

Charron, D.J. and McDevitt, H.O. (1979), *Proc. Natl. Acad. Sci. U.S.A.*, **76**, 6567–6571.

Cohen, C.J. and Murphey-Corb, M. (1983), *Nature (London)*, **301**, 129–133.

Cotner, T., Mashimo, H., Kung, P.C., Goldstein, G. and Strominger, J.L. (1981), *Proc. Natl. Acad. Sci. U.S.A.*, **78**, 3858–3862.

Day, C.E. and Jones, P.P. (1983), *Nature (London)*, **302**, 157–159.

de Kretser, T.A., Crumpton, M.J., Bodmer, J.G. and Bodmer, W.F. (1982a), *Eur. J. Immunol.*, **12**, 214–221.

de Kretser, T.A., Crumpton, M.J., Bodmer, J.G. and Bodmer, W.F. (1982b), *Eur. J. Immunol.*, **12**, 600–606.

Dick, H. (1982), *Immunol. Today* **3**, 200.

Dover, G. (1982), *Nature (London)*, **299**, 111–117.

Dupont, B., Pollack, M.S., Levine, L.S., O'Neill, G.J., Hawkins, B.R. and New, H.T. (1980) in *Histocompatibility Testing, 1980* (P.I. Terasaki, ed.), UCLA Tissue Typing Laboratory, Los Angeles, pp. 693–706.

Fleischnick, E., Raum, D., Alosco, S.M., Gerald, P.S., Yunis, E.J., Awdeh, Z.L., Granados, J., Crigler, J.F., Giles, C.M. and Alper, C.A. (1983), *Lancet*, **i**, 152–156.

Franke, U. and Pellegrino, M.A. (1977), *Proc. Natl. Acad. Sci. U.S.A.*, **74**, 1147–1151.

Gally, J.A. and Edelman, G.M. (1972), *Annu. Rev. Genet.*, **6**, 1–46.

Goodenow, R.S., Stroynowski, I., McMillan, M., Nicolson, M., Eakle, K., Sher, B.T., Davidson, N. and Hood, L. (1983), *Nature (London)*, **301**, 388–394.

Goodfellow, P.N., Jones, E.A., van Heyningen, V., Solomon, E., Bobrow, M., Miggiano, V. and Bodmer, W.F. (1975), *Nature (London)*, **254**, 267–269.

Goodfellow, P.N., Banting, G., Trowsdale, J., Chambers, S. and Solomon, E. (1982), *Proc. Natl. Acad. Sci. U.S.A.*, **79**, 1190–1194.

Goyert, S.M., Shirley, J.E. and Silver, J. (1982), *J. Exp. Med.*, **156**, 550–566.

Gustafsson, K., Bill, P., Larhammar, D., Wiman, K., Claesson, L., Schenning, L., Servenius, B., Sundelin, J., Rask, L. and Peterson, P.A. (1982), *Scand. J. Immunol.*, **16**, 303–308.

Guy, K. and van Heyningen, V. (1982), *Immunol. Today*, **3**, 237.

Hiramatsu, K., Ochi, A., Miyatani, S., Segawa, A. and Tada, T. (1982), *Nature (London)*, **296**, 666–668.

Hood, L., Steinmetz, M. and Goodenow, R. (1982), *Cell*, **28**, 685–687.

Hurley, C., Shaw, S., Nadler, L., Schlossman, S. and Capra, J.D. (1982), *J. Exp. Med.*, **156**, 1557–1562.

Jensenius, J.C. and Williams, A.F. (1982), *Nature (London),* **300,** 583–588.

Kavathas, P., DeMars, R., Bach, F.H. and Shaw, S. (1981), *Nature (London),* **293,** 747–749.

Klein, J. (1979), *Science,* **203,** 516–521.

Klein, J. and Nagy, Z.A. (1982), *Nature (London),* **300,** 12–13.

Knowles, R.W. and Bodmer, W.F. (1982), *Eur. J. Immunol.* **12,** 676–681.

Koch, N., Koch, S. and Hammerling, G.J. (1982), *Nature (London),* **299,** 644–645.

Korman, A.J., Knudsen, P.J., Kaufman, J.F. and Strominger, J.L. (1982), *Proc. Natl. Acad. Sci. U.S.A.,* **79,** 1844–1848.

Kratzin, H., Yang, C., Gotz, H., Pauly, E., Kolbel, S., Egert, G., Thinnes, F.P., Wernet, P., Allevogt, P. and Hilschmann, N. (1981), *Hoppe Seyler's Z. Physiol. Chem.,* **362,** 1665–1669.

Larhammar, D., Wiman, K., Schenning, L., Claesson, L., Gustafsson, K., Peterson, P.A. and Rask, L. (1981), *Scand. J. Immunol.,* **14,** 614–622.

Larhammar, D., Schenning, L., Gustafsson, K., Wiman, K., Claesson, L., Rask, L. and Peterson, P.A. (1982a), *Proc. Natl. Acad. Sci. U.S.A.,* **79,** 3687–3691.

Larhammar, D., Gustafsson, K., Claesson, L., Bill, P., Wiman, K., Schenning, L., Sundelin, J., Widmark, E., Peterson, P.A. and Rask, L. (1982b), *Cell,* **30,** 153–161.

Lee, J.S., Trowsdale, J. and Bodmer, W.F. (1980), *J. Exp. Med.,* **152,** 3s–10s.

Lee, J.S., Trowsdale, J. and Bodmer, W.F. (1982a), *Proc. Natl. Acad. Sci. U.S.A.,* **79,** 545–549.

Lee, J.S., Trowsdale, J., Travers, P., Carey, J., Grosveld, F., Jenkins, J. and Bodmer, W. (1982b), *Nature (London),* **299,** 750–752.

Long, E.O., Wake, C.T., Strubin, M., Gross, N., Accolla, R.S., Carrel, S. and Mach, B. (1982), *Proc. Natl. Acad. Sci. U.S.A.,* **79,** 7465–7469.

Long, E.O., Strubin, M., Wake, C.T., Gross, N., Carrel, S., Goodfellow, P., Accolla, R.S. and Mach, B. (1983a), *Proc. Natl. Acad. Sci. U.S.A.,* (in press).

Long, E.O., Wake, C.T., Gorski, J. and Mach, B. (1983b), *EMBO Journal,* **2,** 389–394.

Malissen, M., Malissen, B. and Jordan, B.R. (1982a), *Proc. Natl. Acad. Sci. U.S.A.,* **79,** 893–897.

Malissen, M., Damotte, M., Birnbaum, D., Trucy, J. and Jordan, B.R. (1982b) *Gene,* **20,** 485–489.

McKean, D.J. (1983), *J. Immunol.,* **130,** 1268–1273.

Mellor, A.L., Golden, L., Weiss, E., Bullman, H., Hurst, J., Simpson, E., James, R.F.L., Townsend, A.R.M., Taylor, P.M., Schmidt, W., Ferluga, J., Leben, L., Santamaria, M., Atfield, G., Festenstein, H. and Flavell, R.A. (1982), *Nature (London),* **298,** 529–534.

Meruelo, D., Offer, M. and Rossomando, A. (1982), *Proc. Natl. Acad. Sci. U.S.A.,* **79,** 7460–7464.

Michaelson, J., Rothenberg, E. and Boyse, E.A. (1980), *Immunogenetics,* **11,** 93–95.

Monaco, J.J. and McDevitt, H.O. (1982), *Proc. Natl. Acad. Sci. U.S.A.,* **79,** 3001–3005.

Murphy, D.B. (1981), in *The Role of the Major Histocompatibility Complex in Immunobiology* (M.E. Dorf, ed.), Garland SPTM, New York, pp. 1–32.

Nathenson, S.G., Uehara, H., Ewenstein, B.M., Kindt, T.J. and Coligan, J.E. (1981), *Annu. Rev. Biochem.,* **50,** 1025–1052.

Nixon, D.F., Ting, J. P–Y. and Frelinger, J.A. (1982), *Immunol. Today,* **3,** 339–342.
Orr, H.T., Lopez de Castro, J.A., Parham, P., Ploegh, H.L. and Strominger, J.L. (1979), *Proc. Natl. Acad. Sci. U.S.A.,* **76,** 4395–4399.
Orr, H.T., Bach, F.H., Ploegh, H.L., Strominger, J.L., Kavanthas, P. and deMars, R. (1982), *Nature (London),* **296,** 454–456.
Parham, P. and Strominger, J. (eds) (1982), *Histocompatibility Antigens, Receptors and Recognition, Series B,* Vol. 14, Chapman and Hall, London.
Parnes, J.R. and Seidman, J.G. (1982), *Cell,* **29,** 661–669.
Parnes, J.R., Valen, B., Felsenfeld, A., Ramanathan, L., Ferrini, U., Appella, E. and Seidman, J.G. (1981), *Proc. Natl. Acad. Sci. U.S.A.,* **78,** 2253–2257.
Pease, L.R., Schulze, D.H., Pfaffenbach, G.M. and Nathenson, S.G. (1983), *Proc. Natl. Acad. Sci. U.S.A.,* **80,** 242–246.
Peterson, P.A., Cunningham, B.A., Berggard, I. and Edelman, G.M. (1972), *Proc. Natl. Acad. Sci. U.S.A.,* **69,** 1697–1701.
Ploegh, H.L., Orr, H.T. and Strominger, J.L. (1980), *Proc. Natl. Acad. Sci. U.S.A.,* **77,** 6081–6085.
Ploegh, H.L., Orr, H.T. and Strominger, J.L. (1981), *Cell,* **24,** 287–299.
Porter, R.R. (1979), in *International Review of Biochemistry* (E.S. Lennox, ed.), University Park Press, Baltimore, pp. 177–212.
Raum, D., Donaldson, V.H., Rosen, F.S. and Alper, C.A. (1980), *Curr. Top. Haematol.,* **3,** 111–174.
Stetler, D., Das, H., Nunberg, J.H., Saiki, R., Sheng-Dong, R., Mullis, K.B., Weissman, S.M and Erlich, H.A. (1982), *Proc. Natl. Acad. Sci. U.S.A.,* **79,** 5966–5970.
Shackelford, D.A. and Strominger, J.L. (1980), *J. Exp. Med.* **151,** 144–165.
Shackelford, D.A., Kaufman, J.F., Korman, A.J. and Strominger, J.L. (1982), *Immunol. Rev.,* **66,** 133–187.
Shaw, S., Kavathas, P., Pollack, M.S., Charmot, D. and Mawas, C. (1981), *Nature (London),* **293,** 745–747.
Shreffler, D.C. (1982), in *Histocompatibility Antigens, Receptors and Recognition, Series B,* Vol. 14 (P. Parham and J. Strominger, eds), Chapman and Hall, London.
Sood, A.K., Pereira, D. and Weissman, S.M. (1981), *Proc. Natl. Acad. Sci. U.S.A.,* **78,** 616–620.
Steinmetz, M., Frelinger, J.G., Fisher, D., Hunkapiller, T., Pereira, D., Weissman, S.M., Uehara, H., Nathenson, S. and Hood, L. (1981a), *Cell,* **24,** 125–134.
Steinmetz, M., Moore, K.W., Frelinger, J.G., Sher, B.T., Shen, F.W., Boyse, E.A. and Hood, L. (1981b), *Cell,* **25,** 683–692.
Steinmetz, M., Minard, K., Horvath, S., McNicholas, J., Frelinger, J., Wake, C., Long, E., Mach, B. and Hood, L. (1982a), *Nature (London),* **300,** 35–42.
Steinmetz, M., Winoto, A., Minard, K. and Hood, L. (1982b), *Cell,* **28,** 489–498.
Strominger, J., *et al.* (1981), in *The Role of the Major Histocompatibility Complex in Immunobiology* (M.E. Dorf, ed.), Garland SPTM, New York, pp. 115–172.
Suggs, S.V., Wallace, R.B., Hirose, T., Kawashima, E.H. and Itakura, K. (1981), *Proc. Natl. Acad. Sci. U.S.A.,* **78,** 6613–6617.
Tragardh, L., Rask, L., Wiman, K., Fohlman, J. and Peterson, P.A. (1979), *Proc. Natl. Acad. Sci. U.S.A.,* **76,** 5839–5842.
Tragardh, L., Rask, L., Wiman, K., Fohlman, J. and Peterson, P.A. (1980), *Proc. Natl. Acad. Sci. U.S.A.,* **77,** 1129–1133.

Trowsdale, J., Lee, J.S., Carey, J., Grosveld, F., Bodmer, J. and Bodmer, W. (1983a), *Proc. Natl. Acad. Sci. U.S.A.*, **80**, 1972–1976.

Trowsdale, J., Lee, J.S. and McMichael, A. (1983b), *Immunol. Today*, **4**, 31–35.

Wake, C.T., Long, E.O. and Mach, B. (1982a), *Nature (London)*, **300**, 372–374.

Wake, C.T., Long, E.O. Strubin, M., Gross, N., Accolla, R., Carrel, S. and Mach, B. (1982b), *Proc. Natl. Acad. Sci. U.S.A.*, **79**, 6979–6983.

Walsh, F.S. and Crumpton, M.J. (1977), *Nature (London)*, **269**, 307–311.

Weiss, E., Golden, L., Zakut, R., Mellor, A., Fahmer, K., Kvist, S. and Flavell, R.A. (1983a), *EMBO Journal*, **2**, 453–462.

Weiss, E.H., Mellor, A., Golden, L., Fahmer, K., Simpson, E., Hurst, J. and Flavell, R.A. (1983b), *Nature (London)*, **301**, 671–674.

Whitehead, A.S., Solomon, E., Chambers, S., Bodmer, W.F., Povey, S. and Fey, G. (1982), *Proc. Natl. Acad. Sci. U.S.A.*, **79**, 5021–5025.

Wiman, K., Larhammar, D., Claesson, L., Gustafsson, K., Schenning, L., Bill, P., Bohme, J., Denaro, M., Dobberstein, B., Hammerling, U., Kvist, S., Servenius, B., Sundelin, J., Peterson, P.A. and Rask, L. (1982), *Proc. Natl. Acad. Sci. U.S.A.*, **79**, 1703–1707.

Womack, J.E., Yan, D.L.S. and Potier, M. (1981), *Science*, **212**, 63–65.

Woods, D.E., Markham, A.F., Ricker, A.T., Goldberger, G., and Colten, H.R. (1982), *Proc. Natl. Acad. Sci. U.S.A.*, **79**, 5661–5665.

Ziegler, A. and Milstein, C. (1979), *Nature (London)*, **279**, 243–244.

Note added in proof

Since writing this review in early 1983 there has been considerable development in the molecular biology of the HLA region. The following references contain some of the new data and serve as sources for further reading.

Class I genes

Lalanne, J.-L., Cochet, M., Kummer, A.-M., Gachelin, G. and Kourilsky, P. (1983), *Proc. Natl. Acad. Sci. U.S.A.*, **80**, 7561–7565.

Brickell, P.M., Latchman, D.S., Murphy, D., Willison, K. and Rigby, P.W.J. (1983), *Nature (London)*, **306**, 756–760.

Class II genes

Larhammer, P., Hyldig-Nielson, J.J., Servenius, B., Andersson, G., Rask, L. and Peterson, P.A. (1983), *Proc. Natl. Acad. Sci. U.S.A.*, **80**, 7313–7317.

Chang, H.C., Moriuchi, T. and Silver, J. (1983), *Nature (London)*, **305**, 813–815.

Roux-Dosseto, M., Auffray, C., Lillie, J.W., Boss, J.M., Cohen, D., De Mars, R., Mawas, C., Seidman, J.G. and Strominger, J.L. (1983), *Proc. Natl. Acad. Sci. U.S.A.*, **80**, 6036–6040.

Class III genes

Carroll, M.C., Campbell, D., Bentley, D.R. and Porter, R.R. (1984), *Nature (London)*, **307**, 237–241.

General

Steinmetz, M. and Hood, L. (1983), *Science*, **222**, 723–733.

5 Cell Genetic Analysis of the Receptor Systems for Bioactive Polypeptides

NOBUYOSHI SHIMIZU

EDITOR'S INTRODUCTION

Antibodies are not the only reagents which can be used to probe the surface of cells. Bioactive peptides, hormones and growth factors also bind to the cell surface. In this latter case the molecules which bind by definition recognize functionally important cell surface molecules. Genetic analysis has been used to map the genes controlling growth factor receptors and to elucidate the mechanism by which growth factors mediate these biological effects. In Chapter 5 Nobuyoshi Shimizu describes several novel approaches to studying receptors for growth factors. Particularly interesting is the coupling of toxic molecules to growth factors to select cell variants which either do not bind growth factors or fail to correctly process the internalized growth factors.

Acknowledgements

The author thanks W.K. Miskimins, Y. Shimizu, M.A. Behzadian, I. Kondo, S. Gamou, F. Ito, J. Lewis and J. Hunts for their contributions to the research projects presented in this chapter. The research was supported by NIH grant GM-24375, ACS Grant JFA-9 and US–Japan Cooperative Cancer Research Program Grant.

Genetic Analysis of the Cell Surface
(*Receptors and Recognition,* Series B, Volume 16)
Edited by P. Goodfellow
Published in 1984 by Chapman and Hall, 11 New Fetter Lane, London EC4P 4EE

5.1 INTRODUCTION

Polypeptide hormones and growth factors are among the bioactive polypeptides that play an important role in the regulation of mammalian cell growth and differentiation. Recently, it has become a central theme in biology to understand the mechanisms by which these bioactive polypeptides influence the morphology, metabolism and gene expression of cells. A complete understanding of these mechanisms will eventually provide a basis for unveiling the causes of many complex diseases, such as cancer, which apparently result from abnormalities in cellular functions.

Insulin and epidermal growth factor (EGF) are two well-characterized bioactive polypeptides whose biochemical and structural properties are well known. Although insulin has a known distinct role in hormonal action, both of these polypeptides are potent mitogens to cultured cells. Furthermore, insulin and EGF apparently act through similar though not identical mechanisms.

There are a number of other bioactive polypeptides that are functionally and structurally related to insulin and EGF. These are the insulin-like growth factors (IGF-I and IGF-II), somatomedins, multiplication stimulating activity (MSA), nerve growth factor (NGF), fibroblast growth factor (FGF) and the EGF-like growth factors which include various types of transforming growth factors (TGFs). There are also several other bioactive polypeptides present in serum that are involved in cell growth control. These include α_2-macroglobulin, thrombin, transferrin and low-density lipoprotein (LDL). Considerable interest has been focused on insulin, EGF and LDL, and these polypeptides are serving as model systems for exploring the mechanisms of action of bioactive polypeptides.

Most of our current understanding of the molecular action of these polypeptides is derived from studies using biochemical, pharmacological and immunological methods. These studies have provided a general scheme of molecular action: the bioactive polypeptides initially interact with specific cell surface receptors and this binding elicits a signal(s) to the cells, which then undergo numerous early and delayed biochemical reactions. However, it has not been possible to construct a unified concept as to how they regulate cellular processes.

This chapter describes the current status of the knowledge of the molecular action of bioactive polypeptides and then reviews an emerging approach, based on the concepts and techniques of somatic cell genetics, which allows a better understanding of the mechanism of action of these polypeptides. The strategy used is to select genetic variants deficient in some aspects of cellular responsiveness to these bioactive polypeptides. These cell variants can then be analysed at genetic and molecular levels for these lesions. The availability of permanent cell lines that exhibit specific responses to each of these bioactive polypeptides in a stable fashion has made this genetic analysis feasible. It is particularly noteworthy that application of hormone–toxin hybrid proteins has provided cell variants which are extraordinary in the nature of their receptors and post-receptor pathways. Although the genetic approach to bioactive polypeptide receptor action is still in its infancy, it has already provided experimental systems by which regulation of cell growth is studied as well as providing evidence for determining the chromosomal localization of the receptor structural genes.

5.2 A COMMENTARY ON THE MECHANISMS OF BIOACTIVE POLYPEPTIDE ACTION

5.2.1 Receptors

The initial interaction of both insulin and EGF with cells is binding to specific cell surface receptors. This binding is rapid, saturable and dependent on time, temperature and pH (Roth *et al.*, 1979; Carpenter and Cohen, 1976). The insulin receptor has been found to be a multi-subunit glycoprotein (Jacobs *et al.*, 1979). It consists of two α-subunits of M_r = 130 K and two β-subunits of M_r = 90 K which are linked covalently by disulphide bonds in a manner similar to immunoglobulins (Pilch and Czech, 1980; Massague *et al.*, 1980; Jacobs and Cuatrecasas, 1981). The α-subunit apparently contains the insulin-binding site and the β-subunit is susceptible to a specific proteolytic cleavage, producing a M_r = 45 K fragment (Massague *et al.*, 1980, 1981b).

There are two types of IGFs (Blundell *et al.*, 1983). Receptors for IGF-I have an $\alpha_2\beta_2$ subunit structure similar to that of insulin receptors and have a weak affinity for binding insulin and IGF-II (Massague and Czech, 1982; Kasuga *et al.*, 1981, 1982b). Receptors for IGF-II have a single subunit of M_r = 260 K and are similar to MSA receptors (Massague and Czech, 1982; Massague *et al.*, 1981a). Insulin does not compete with MSA for its binding to the receptors (Nagarajan *et al.*, 1982).

The EGF receptor has been identified as a glycoprotein of M_r = 170 K (Das *et al.*, 1977; Das and Fox, 1978; Carpenter *et al.*, 1979; Hock *et al.*, 1979; Baker *et al.*, 1979). It apparently consists of a single subunit, but may exist as a multimer in its native cell membrane. Interestingly, the EGF receptor is also sensitive to a proteolytic cleavage, producing a M_r = 150 K receptor species (Linsley and Fox, 1980; Cohen *et al.*, 1982b; Gates and King, 1982). Receptors for the TGFs, NGF and others have been characterized (Czech *et al.*, 1983; Massague *et al.*, 1982b).

5.2.2 Receptor–ligand binding

Binding of both EGF and insulin to their respective receptors is a complex interaction as shown in curvilinear Scatchard plots (Bylund, 1980). Such results have been interpreted as an indication of the existence of multiple binding sites with varying affinities for the ligand or as a negative co-operativity phenomenon. In the case of insulin the negative co-operativity model is supported by the finding that unlabelled insulin stimulates the dissociation of prebound labelled insulin (DeMeyts *et al.*, 1973). The multiple site model for insulin binding is supported by the ability of insulin to bind with lower affinity to IGF-I receptors (Rechler *et al.*, 1980). There is no evidence for negative co-operativity in EGF binding and the curvilinear Scatchard plot is thought to be due to the presence of high- and low-affinity receptors (King and Cuatrecasas, 1982). It is believed that ligand binding to the high-affinity receptors is responsible for the biological effects of these polypeptide ligands (Schechter *et al.*, 1978) and that low-affinity receptors regulate external ligand levels through a lysosomal degradation pathway. Whether these receptors are distinct gene products or interchangeable forms of the same gene product is currently unknown.

5.2.3 Cellular responses

It was 1924 when insulin was first added to cultured cells to stimulate their growth (Gey and Thalhimer, 1924). In more recent times, insulin is almost invariably added to serum-free hormone-defined media (Hayashi *et al.*, 1978; Barnes and Sato, 1980; Sato *et al.*, 1982). In general, for stimulation of cell growth, insulin is required at concentrations 100–1000 times that required for stimulation of amino acid uptake (0.1 nM). This growth-stimulatory effect of insulin is proposed to be through the receptors for insulin-like growth factors such as IGFs and MSA rather than through the insulin receptors (Barnes and Sato, 1980; King *et al.*, 1980b). However, cell lines like H35 rat hepatoma and some human mammary tumour cells are stimulated by 0.1 nM insulin (Osborne *et al.*, 1978; Allegra and Lippman, 1978; Barnes and Sato, 1979; Koontz and Iwahashi, 1981; Massague *et al.*, 1982a; Czech *et al.*, 1983), so it is likely that, at least for some cell types, insulin stimulates cell growth through the insulin receptors.

It is postulated that target cells whose DNA synthesis is stimulated by insulin alone are likely to be already competent and limited principally for the stimulation process regulated by insulin (Stiles *et al.*, 1979). Typically, insulin acts synergistically with other growth factors to promote proliferation (DeAsua *et al.*, 1977a,b; Sato and Ross, 1979). It has recently been shown that insulin rapidly modulates the apparent affinity of the IGF-II receptors (Oppenheimer *et al.*, 1983).

The early effects of these bioactive polypeptides which appear within seconds or minutes include stimulation of membrane transport of sugars and amino acids, stimulation of ion fluxes, and stimulation of enzymes involved in glycolysis, lipid metabolism and glycogen synthesis. The delayed effects which appear several hours after ligand binding include the stimulation of protein, RNA and DNA synthesis and cell division.

It is still an open question whether binding of these ligands to their surface receptors is enough to produce their biological effects or whether further intracellular processing of the ligands or their receptors is necessary. Evidence suggests that both are necessary for production of a 'full' mitogenic response.

5.2.4 Proposed signals

Numerous studies have proposed signals which correlate with the early or delayed cellular effects. These signals include a glucose transport system, Ca^{2+} ions and diacylglycerol as second messengers, chemical mediators, phosphorylation reactions, polyamines, ADP-ribosylation reactions and cytoskeletal changes.

In the case of membrane transport, it is possible that the ligand–receptor complex interacts either directly or through an effector molecule with the transport system on the plasma membrane. Direct interaction was suggested by the finding that both membrane fluidity and receptor mobility increase upon binding of the ligands (Schlessinger *et al.*, 1978a,b; Jacobs and Cuatrecasas, 1981). However, recent evidence suggests the existence of 'glucose transporters' which are rapidly mobilized from an intracellular location to the cell surface upon stimulation by insulin (Suzuki and Kono, 1980; Cushman and Wardzala, 1980; Kono *et al.*, 1982).

Another possible mechanism which has been considered for the activation of the early effects of insulin and EGF is the formation of a second messenger. This idea is based on the analogy to the mechanism of action of glucagon and β-adrenergic hormones (Rodbell *et al.*, 1971; Lefkowitz *et al.*, 1973). These hormones bind to cell surface receptors and, through an effector, activate adenylate cyclase and increase the production of cyclic AMP (cAMP), the second messenger (cAMP) then interacts with specific protein kinases which in turn regulate the activities of target enzymes involved in the metabolic responses (Glass and Krebs, 1980; Cohen, 1982). Attempts to correlate the level of cyclic nucleotides with insulin and EGF responses have been unsuccessful and their postulated roles as second messengers appear unlikely. Ion fluxes have also been proposed as second messengers of insulin and EGF action since ion fluxes may cause membrane hyperpolarization which could result in perturbations in the membrane organization. In this regard, Ca^{2+} ions are considered potent second messengers since they bind with high affinity to a specific intracellular protein, calmodulin. A number of cellular activities have been attributed to calmodulin including activation of a specific protein kinase (Wang and Weissman, 1979; Cheung, 1980; Means and Dedman, 1980). The involvement of Ca^{2+} in the mechanisms of cell growth stimulation by EGF and insulin has recently been suggested (Miskimins and Shimizu, 1983; Goewert *et al.*, 1983; McKeehan *et al.*, 1982). Furthermore, calmodulin has been shown to be an important element for cells to progress through the cell cycle (Chafouleas *et al.*, 1982). The level of this protein doubles in late G1- or S-phase.

Recently, a chemical mediator of insulin action has been identified (Jarett and Seals, 1979; Jarett *et al.*, 1981; Larner *et al.*, 1981). This mediator is a heat-stable small protein of M_r = 1000–3000 and appears to be formed by a proteolytic event following insulin binding to its receptor. The mediator protein isolated from adipocytes can stimulate mitochondrial pyruvate dehydrogenase as well as glycogen synthase phosphatase. It can be generated in a cell-free system composed of adipocyte cell membrane and mitochondria upon addition of insulin (Seals *et al.*, 1979a,b; Seals and Czech, 1980). Another chemical mediator of small peptide nature has been detected in rat liver which can regulate pyruvate dehydrogenase and adenylate cyclase as well as acetyl-CoA carboxylase activity (Saltiel *et al.*, 1982, 1983).

An additional mechanism of signal transduction by cells stimulated with EGF and insulin may be through phosphorylation–dephosphorylation reactions. It is well known that many enzymes associated with insulin's hormonal action in adipocytes are regulated by the state of phosphorylation level (Cohen, 1982). The recent discovery that both insulin and EGF rapidly stimulate phosphorylation of specific cell membrane-associated proteins led to the conclusion that a primary action of these polypeptides is the stimulation of specific protein kinase activities (Seals *et al.*, 1979a,b; Carpenter *et al.*, 1978). In fact, it has recently been shown that both insulin and EGF stimulate phosphorylation of their receptors (Cohen *et al.*, 1980; Kasuga *et al.*, 1982a). Furthermore, evidence strongly suggests that both the insulin receptor and EGF receptor are themselves protein kinases (Cohen *et al.*, 1980, 1982a; Kasuga *et al.*, 1982c). In the case of the insulin receptor, the larger α-subunit retains the kinase domain (ATP binding site) and the smaller β-subunit is susceptible to phosphorylation (Kasuga *et al.*, 1982a), whereas the EGF receptor subunit appears to contain both the kinase domain and the EGF binding domain in the same molecule (Cohen *et al.*, 1980, 1982a). It should be noted that these receptors are phosphoproteins having their phosphorylation sites on serine and threonine residues while the ligand-stimulated phosphorylation takes place on tyrosine residues (Ushiro and Cohen, 1980; Hunter and Cooper, 1981; Kasuga *et al.*, 1982a).

This tyrosine phosphorylation appears common to several other bioactive polypeptide systems including platelet-derived growth factor (PDGF) (Heldin and Rönnstrand, 1983) and somatomedin C (Jacobs *et al.*, 1983) and has an important biological implication in terms of cellular transformation since the transforming proteins of numerous RNA tumour viruses are found to be tyrosine-specific protein kinases (Cooper and Hunter, 1981). A close relationship between the two kinases, the receptor-associated kinase and the viral oncogene-associated kinase, has been demonstrated in several different ways. For example, antibodies to Rous sarcoma virus transforming protein pp60 sarc are phosphorylated on tyrosine by the EGF receptor kinase (Chinkers and Cohen, 1981; Kudlow *et al.*, 1981) and synthetic oligopeptides, such as angiotensin-II, which contain tyrosine residues are equally phosphorylated by both kinases (Pike *et al.*, 1982). This finding is particularly interesting in the light of the fact that cellular homologues to the transforming genes exist in uninfected normal cells (Cooper, 1982). The most intriguing recent finding is that the platelet-derived growth factor (PDGF) gene has a similar sequence to that of the *sis* oncogene (Waterfield *et al.*, 1983). These areas are currently under intensive study. It should, however, be mentioned at this point that while receptor phosphorylation may be a necessary signal for mitogenic stimulation, it may not be sufficient for this since both CNBr-modified EGF and IgG antibodies against the EGF receptor stimulate membrane phosphorylation but not DNA synthesis (Schreiber *et al.*, 1981a,b).

An additional phosphorylation reaction stimulated by both EGF and insulin is that of the ribosomal protein S6 (Smith *et al.*, 1979). This phosphorylation, however, is not on tyrosine residues. Recent evidence suggests that S6 phosphorylation enhances polysome formation, thereby increasing the rate of protein synthesis (Thomas *et al.*, 1982). S6 phosphorylation is synergistically stimulated by either EGF and insulin, EGF and PGF 2α, or insulin and PDGF but is apparently not a sufficient signal for DNA synthesis in mitogen-stimulated cells (Nilsen-Hamilton *et al.*, 1982; Chambard *et al.*, 1983).

Recently, diacylglycerol has been shown to act as a second messenger in transmembrane control of intracellular events (Takai *et al.*, 1979; Nishizuka and Takai, 1981). The finding is that when platelets are treated with thrombin, a phospholipase C activity is stimulated. This leads to an increase in phosphatidylinositol turnover resulting in the formation of diacylglycerol. Diacylglycerol then activates a specific protein kinase, C-kinase, which is absolutely dependent on Ca^{2+}. This C-kinase is known to specifically phosphorylate a protein of $M_r = 40$ K, the exact function of which is at present unknown. It is, however, noted that C-kinase stimulates phosphorylation on serine but not on threonine or on tyrosine (Sano *et al.*, 1983). It has been shown that a tumour-promoting agent 12-*O*-tetradecanoyl phorbol 13-acetate (TPA) can replace diacylglycerol and directly activate C-kinase (Castagna *et al.*, 1982). The findings that TPA modulates EGF and insulin receptors (Lee and Weinstein, 1978; Shoyab *et al.*, 1979; Brown *et al.*, 1979; Magun *et al.*, 1980; Thomopoulos *et al.*, 1982) and has many growth factor-like effects on cultured cells (Weinstein *et al.*, 1979; Mastro, 1982) and that insulin is known to affect phospholipid turnover (Fain, 1980) suggest a role for C-kinase in the actions of EGF and insulin.

Polyamines have also been shown to be related to cell growth. In cultured cells the synthesis of polyamines is regulated by the enzyme ornithine decarboxylase (ODC). EGF and insulin stimulate the activity of ODC (Stastny and Cohen, 1970; Tomita *et al.*, 1981); thus polyamines are involved at certain steps of the signal transduction stimulated by these bioactive polypeptides.

EGF and insulin may act at the level of chromatin structure since it has been shown that these polypeptides induce poly ADP-ribosylation in growth-arrested mouse fibroblasts in culture and that this induction is coincident with the entry into DNA synthesis in S-phase (Shimizu and Shimizu, 1981). EGF-enhanced acetylation of nuclear proteins has also been reported (Kaneko, 1983). Thus, alterations in chromatin structure caused by modification of nuclear proteins may be an important event in the initiation of DNA synthesis. In this regard it has recently been shown that EGF induces DNA topoisomerases which assemble with several other enzymes to form a DNA-synthesizing complex and that the complex moves from the cellular cytosol to the nucleus to initiate DNA synthesis (Miskimins *et al.*, 1983). Interestingly, the activity of DNA topoisomerase has been found to be regulated by ADP-ribosylation (Ferro *et al.*, 1983).

The production of a mitogenic signal by EGF and insulin is also apparently related to alterations in the cellular cytoskeleton. Colchicine and other microtubule disrupters have been shown to enhance the ability of EGF and insulin to stimulate cellular DNA synthesis (Otto, 1982). On the other hand, taxol and dihydrocytochalasin B inhibit the ability of these polypeptides to induce DNA synthesis (Schiff *et al.*, 1979; Crossin and Carney, 1981; Maness and Walsh, 1982). These drugs are inhibitors of microtubule depolymerization and actin polymerization, respectively. Although drug effects must be interpreted with caution, the cytoskeleton's constituents appear to be involved in transmitting a mitogenic signal.

Using the hybridoma technique, a number of monoclonal antibodies have recently been produced which react with EGF receptors. Use of these monoclonal anti-receptor antibodies has provided new evidence that the biological information for cell growth stimulation may reside in the receptor molecules. The monoclonal anti-receptor antibody of IgM nature caused the induction of early and delayed biological responses when added to the cells (Schreiber *et al.*, 1981a, 1983). In contrast, neither the anti-receptor antibodies of IgG nature nor Fab fragments of the IgM antibody generated any biological effects of EGF (Schreiber *et al.*, 1983; Waterfield *et al.*, 1982). However, upon addition of a second antibody recognizing the initial IgG antibody or Fab fragments the cellular DNA synthesis was significantly enhanced (Shreiber *et al.*, 1981a, 1983). It is postulated that these differences were related to the valency of the antibody and the possible role of receptor clustering in the generation of mitogenic signal (Schreiber *et al.*, 1981a).

5.2.5 Internalization and intracellular processing

Numerous recent studies have demonstrated that EGF, insulin and many other cell surface binding proteins are internalized and delivered to lysosomes where they are degraded (Goldstein *et al.*, 1979; Pastan and Willingham, 1981a,b). Electron microscopy, EM-radioautography, fluorescent microscopy, specific inhibitors and cell fractionation have all been applied to the study of the pathway of this receptor-mediated endocytosis of the ligand (Anderson *et al.*, 1977; Goldfine, 1981a; Gorden *et al.*, 1978; Bergeron *et al.*, 1979; Haigler *et al.*, 1978, 1979, 1980a,b; Wall *et al.*, 1980; Willingham *et al.*, 1983; Schlessinger *et al.*, 1978a,b; Carpenter and Cohen, 1976; Marshall and Olefsky, 1979; King *et al.*, 1980a; Shimizu and Shimizu, 1980; Suzuki and Kono, 1980; Fine *et al.*, 1981). The receptor–ligand complexes rapidly cluster into clathrin-coated pits in the plasma membrane. The coated pits pinch off from the membrane to form coated vesicles (endosomes) which then serve to transport the receptor–ligand complex to lysosomes. This was best demonstrated for low-density lipoprotein (LDL) (Goldstein *et al.*, 1979). However, recent studies using α_2-macroglobulin and asialoglycoproteins as ligands suggest that the clathrin coat is lost during internalization and that the receptor–ligand complexes are transported intracellularly in uncoated vesicles (Willingham and Pastan, 1982; Wall *et al.*, 1980; Dickson *et al.*, 1981).

These uncoated vesicles have been termed receptosomes (Pastan and Willingham, 1981a,b) and it has been suggested that their function is to avoid early fusion with lysosomes. Furthermore, it has been shown that receptosomes accumulate in the Golgi region (GERL) of the cell before they deliver their contents to lysosomes. A similar pathway has been identified for EGF (Willingham and Pastan, 1982). However, other data indicate that EGF is taken up through both coated and uncoated pits and that EGF-containing endocytic vesicles fuse directly with multivesicular bodies and lysosomes (Haigler *et al.*, 1979; McKanna *et al.*, 1979). In the case of insulin, the pathway of endocytosis looks more complex: it is found in coated vesicles (Gorden *et al.*, 1978, 1982a,b), lysosomes (Gorden *et al.*, 1978, 1982a,b), the Golgi (Bergeron *et al.*, 1979; Posner *et al.*, 1982), endoplasmic reticulum (Goldfine, 1981a,b), nuclei (Goldfine *et al.*, 1977), and an unidentified dense non-lysosomal organelle (Bergeron *et al.*, 1979). The coated vesicles containing insulin have been shown to contain its receptors (Khan *et al.*, 1981, 1982). The exact fate of these receptors has not been elucidated although receptor internalization and recycling phenomena have been demonstrated (Fehlmann *et al.*, 1982, 1983; Marshall *et al.*, 1981; Berhanu *et al.*, 1982; Ronnett *et al.*, 1982a,b; Wang *et al.*, 1983; Brown *et al.*, 1983). Receptor degradation has also been shown for EGF (Das and Fox, 1978).

Our recent study shows that EGF is processed through two distinct endocytic routes which are regulated by the cell depending upon its physiological state (Miskimins and Shimizu, 1982a,b; see Fig. 5.1). The pathway leading to lysosomal association is the one for ligand degradation. In the second pathway, EGF is taken up into a Golgi-like subcellular component, from which a portion of the ligand is translocated into a dense lysosomal enzyme-deficient organelle. Use of the degradation pathway is enhanced by deprivation of serum and amino acids while use of the non-degradative pathway is stimulated in complete growth medium containing serum. The non-lysosomal organelles have

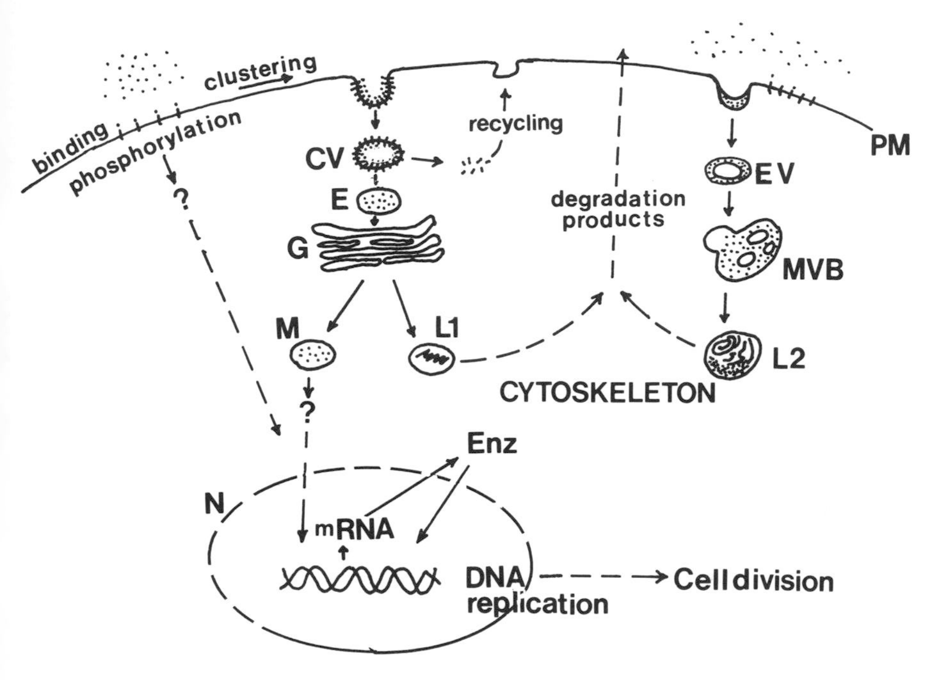

Fig. 5.1 Model for the action of EGF receptor system (modified from Miskimins and Shimizu, 1982a). PM, plasma membrane; CV, coated vesicle; E, endosome; G, Golgi apparatus; M, mitosome; L1, primary lysosome; L2, secondary lysosome; EV, endocytic vesicle; MVB, multivesicular body; N, nucleus; Enz, enzymes. See the text for details.

been termed 'mitosomes' since delivery of the mitogen, in this case EGF, into this component is closely correlated to the ability of that mitogen to induce DNA synthesis via the production of a mitogenic signal (Miskimins and Shimizu, 1983). Mitosomes are distinct from lysosomes in several biochemical criteria (Miskimins and Shimizu, 1983).

There are other hypotheses that insulin and EGF may interact with intracellular binding sites to produce some of their biological effects (Bergeron *et al.*, 1979; Goldfine, 1981a,b; Posner *et al.*, 1978). Specific binding sites, perhaps equivalent to receptors, for insulin have been found in ER, nuclear and Golgi membranes. The receptors found in the former two fractions are different from the plasma membrane receptors in terms of optimal pH and salt concentration for binding and heat stability. Receptors on the Golgi membrane are quite similar to those in the plasma membrane. However, neither the subunit structure of these intracellular receptors nor the genetic relation to their plasma membrane counterparts have been elucidated. Both insulin and EGF have been localized in nuclei following endocytosis (Goldfine *et al.*, 1977;

Johnson *et al.*, 1980), suggesting that these ligands have a direct role in the control of cellular functions. Recent evidence suggests that insulin regulates synthesis of a specific messenger RNA in a cell-free system using isolated nuclei (Purrello *et al.*, 1983).

5.3 ISOLATION OF CELL VARIANTS INVOLVED IN HORMONE RESPONSIVENESS

It is clear from the above discussion that complete understanding of the mechanisms of action of insulin, EGF and other bioactive polypeptides awaits further comprehensive investigations. Knowing the associations of numerous cellular and molecular events with the action of these polypeptides, we and others have made genetic approaches to this important problem. To facilitate genetic studies, attempts have been made to isolate cell variants that are either deficient or extraordinary in hormone response. These cell variants can be found in existing permanent cell lines or they can be isolated from mutagenized cell populations. The selection procedures for hormone-responsiveness variants are distinct from those for the isolation of drug-resistant mutants since polypeptide hormones and growth factors are not only non-toxic to normal mammalian cells in culture, but required for their proliferation. Because multiple gene products are involved in the mechanisms of bioactive polypeptide action, selection of specific variants is more difficult. The variants, once isolated, can be analysed by somatic cell hybridization to determine whether the nature of the defect is genetic or epigenetic and to identify the corresponding gene. In addition, the variants can be analysed biochemically to identify the lesions at the molecular level and to provide knowledge regarding the regulation of the cellular response to these bioactive polypeptides. This section describes various selection procedures and characteristics of the variants isolated by these methods.

5.3.1 Search for spontaneous mutations

Despite the importance of bioactive polypeptides for cell growth, a number of permanent cell lines have been found to lack bioactive polypeptide binding activities. This was particularly noted for EGF binding ability and the cell lines include the mouse L-cell-derived A9 and B82 lines and the Chinese hamster ovary CHO line. Since these cells are morphologically transformed, reduced EGF binding ability due to the sarcoma virus-induced transformation (Todaro *et al.*, 1976, 1979) was suspected. In these RNA virus-transformed cells, sarcoma growth factors are produced which compete with EGF receptors for binding. The lack of EGF binding may be caused by alterations in sugar moieties of the receptor. However, it has been proved that A9 cells lack the receptor proteins (Shimizu *et al.*, 1980a; Behzadian *et al.*, 1982). Furthermore, EGF binding activity was not restored after fusion of A9 cells with B82 cells, indicating no complementation (Davies *et al.*, 1980). Thus, it is likely that A9 cells and B82 cells have the same mutation in the receptor gene. It may be interesting to note that A9 cells are also deficient in MSA receptors but have a little insulin binding activity (Shimizu *et al.*, 1981). Many other cell lines have also been found to be negative for EGF binding, but these cases may be related to the state of differentiation of the cells (Adamson and Rees, 1981). Unlike EGF, an insulin receptor-negative cell line is rare. To our knowledge, the only existing cell line which exhibits no insulin binding is mouse myeloma P3NP (Shimizu and Shimizu, unpublished).

Rat hepatoma H35 cells possess both insulin and IGF-II receptors but lack IGF-I receptors (Massague and Czech, 1982). Mouse 3T3-L1 preadipocytes express receptors for insulin, IGF-I and IGF-II. When they differentiate into adipocytes, the receptors for insulin and IGF-I drastically increase while the number of IGF-II receptors remains constant (Massague and Czech, 1982). Human lymphocytes (RPMI 16666 and RPMI 17666) possess a small number of insulin receptors but these cell lines contain neither IGF-I receptors nor IGF-II receptors (Massague and Czech, 1982). Whether and how these receptors relate to regulation of receptor gene expression is an interesting problem.

Receptor-hyperproducing cells have also been discovered. The human epidermoid carcinoma cell line A431 possesses 2×10^6–3×10^6 EGF receptors per cell (Fabricant *et al.*, 1977). This value is 10–20 times the number of receptors in normal human diploid fibroblasts (Carpenter *et al.*, 1975). A431 cells, however, completely lack receptors for platelet-derived growth factor, PDGF (Glenn *et al.*, 1982). Other EGF receptor-hyperproducing cell lines are human KB and UCVA-1 (Fabricant *et al.*, 1977; Gamou, Kim and Shimizu, unpublished work). Human myeloma IM-9 cells are known to contain 10 times more insulin receptors than diploid fibroblasts, which usually retain 5×10^4 receptors per cell (Van Obberghen *et al.*, 1981). Whether the overproduction is regulated at transcription, translation or other levels is not clear at present. Possible relation to carcinogenesis remains to be elucidated.

There are a number of human fibroblast lines which have been obtained from patients with certain genetic diseases involved in polypeptide receptor functions. Such examples are cells from patients with diabetes and familial hypercholesterolaemia carrying deficiencies in receptor and post-receptor functions (Roth and Grunfeld, 1981; Tolleshaug *et al.*, 1983). Cells from cancer patients also often have abnormalities in receptor function. The reduction in the number of transferrin receptors, for example, appears to be related to certain types of cancer (Trowbridge and Omary, 1981).

5.3.2 Use of mitogenic responses

Variants have been selected from mouse 3T3 fibroblasts for their inability to initiate DNA synthesis and cell division when the quiescent, non-dividing cells are treated with mitogens. The selection is based on the removal of mitogenically responsive cells in the presence of insulin or EGF and vinblastine sulphate (see Fig. 5. 2) (Shimizu and Shimizu, 1980). Among these variants, the IN-2 line has been characterized and found to be unique. IN-2 exhibited temperature-dependent insulin binding. Scatchard analysis of the binding data indicated that the insulin receptor's affinity was substantially reduced: $K_d1 = 1.3\times10^{-9}$ M, $K_d2 = 8.4\times10^{-8}$ M for IN-2 as compared with $K_d1 = 4.4\times10^{-10}$ M, $K_d2 = 3.2\times10^{-9}$ M for parental BALBc/3T3, The rate of

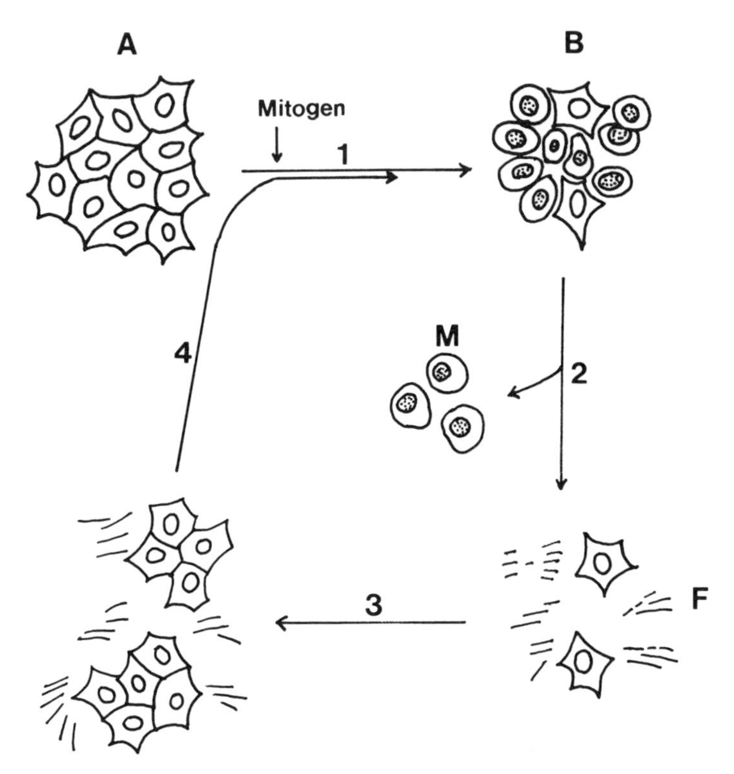

Fig. 5.2 Selection of cell variants using mitogenic response (modified from Shimizu and Shimizu, 1980). A, non-dividing cells at the G0/G1-phase; B, mixed populations of cells at G0/G1- and M-phases; M, mitotic cells; F, cell fragments. Step 1, release from G0/G1-blockage by mitogens such as insulin, EGF and TPA and subsequent mitotic inhibition by vinblastine sulphate; step 2, removal of mitotic cells by mechanical shaking; step 3, stimulation of growth by serum; step 4, repeat the cycle. See the text for details.

insulin dissociation from the receptors was significantly faster in the variant, indicating a low-affinity nature in the receptor. In this variant cell line, insulin interacted with its receptor in a negatively co-operative manner. With respect to receptor structures, it was shown that the gel profiles of the ^{125}I-insulin–receptor complex are not different from those of parental 3T3 (Shimizu *et al.*, unpublished). It was postulated that high affinity and low affinity receptors are likely to be the same gene products and perhaps it is their topology in the membrane that causes different affinities. It is known that in mouse BALBc/3T3 fibroblasts insulin has no effect on glucose transport in the exponentially growing state but it rapidly stimulates glucose transport in the contact-inhibited quiescent state (Bradley and Culp, 1974). In IN-2 cells, insulin was ineffective in accelerating the uptake of 2-deoxy-D-glucose (Shimizu and Shimizu, 1980). As expected from the selection scheme, DNA synthesis in IN-2 cells was also unresponsive to insulin while DNA synthesis in the parental 3T3 cells was significantly stimulated at 5×10^{-11} M. Additional evidence suggests that the insulin bound to the low affinity insulin receptors on IN-2 cells is subject to internalization and lysosomal degradation (Shimizu and Shimizu, 1980). From these results it was concluded that internalization and intracellular processing of insulin cannot, by themselves, be correlated with the mitogenic action of insulin and that high affinity receptors are crucial for these insulin actions. Since the number of low affinity insulin receptors calculated for IN-2 cells is far more than that of MSA receptors in the same cell line (Shimizu, unpublished), these receptors are probably derived from different genes.

Utilizing mitogenic stimulation with EGF and mitogenic arrest with colcemid, several EGF-non-responsive variants have been isolated from Swiss/3T3 cells (Pruss and Herschman, 1977). Among them, NR-6 was found to have completely lost EGF binding capacity due to lack of EGF receptors. However, this line was responsive to other mitogens such as PMA (phorbol myristate acetate), prostaglandin F2α and insulin. It is possible that signals generated by insulin, for example, emerge into a mitogenic post-receptor pathway in common with EGF's. The mutation in the EGF receptor of NR-6 cells appears to be a mis-sense or nonsense mutation rather than a deletion since EGF receptor-positive variants were isolated after mutagenization with ethylmethanesulphonate (Shimizu, unpublished).

Similarly, variants were isolated from Swiss/3T3 cells for their inability to respond to the mitogenic action of TPA (12-*O*-tetradecanoyl phorbol 13-acetate) (Butler-Gralla and Herschman, 1981). TNR-2 and TNR-9 are two such variant lines that no longer respond to TPA. Both variant lines bind [^{3}H]PDBU (phorbol dibutyrate), a radioactive analogue of TPA, and therefore apparently have TPA receptors. The TNR-2 line was incapable of binding EGF and hence non-responsive to EGF while the TNR-9 line was responsive to EGF in a normal way. It was suggested that these variant lines have defects in post-receptor functions of TPA. TPA-resistant variants isolated from JB-6 mouse epidermal cell line (Colburn *et al.*, 1981) fell into three phenotypic classes when their growth ability in soft agar was assayed: (i) anchorage-independent transformants, (ii) variants resistant to promotion of anchorage independence by TPA, and (iii) variants sensitive to promotion by TPA. The existence of the class (iii) variants led these authors to propose that the two events, TPA-induced mutagenesis and late-stage promotion of tumour cell phenotype, can be dissociated.

Dihydroteleocidin B (DHTB) is an indole alkaloid tumour promoter and has a completely different chemical structure from TPA. Both DHTB and TPA stimulate growth-arrested fibroblastic cells from many origins to initiate DNA synthesis and undergo cell division (Weinstein *et al.*, 1979; Rozengurt, 1980; Fujiki *et al.*, 1981). DNA synthesis in resting cells from a pre-adipocyte line, 3T3-L1, was also initiated by both TPA and DHTB and in both cases increased essentially to the same level (Shimizu *et al.*, 1983b). In both 3T3-L1 and the parental Swiss/3T3 cells, the DNA synthesis triggered by DHTB resulted in division of the cells. However, DNA synthesis stimulated by TPA did not facilitate cell division of 3T3-L1 cells. Several variants were selected from 3T3-L1 cells by removing mitogenically responsive cells in the presence of DHTB and vinblastine sulphate (Shimizu *et al.*, 1983a). Two variants, 1-2 and 2-3, were found whose DNA synthesis can not be stimulated by TPA or DHTB. Another variant line, 3-4, was found whose DNA synthesis can be stimulated by DHTB but which can not subsequently undergo cell division. From these findings, it was suggested that the actions of DHTB and TPA are similar in that both can trigger DNA synthesis in G1-arrested 3T3-L1 cells, but only DHTB can further act on 3T3-L1 cells during the G2-phase to stimulate mitosis.

5.3.3 Use of inhibitory effects

The Cloudman melanoma S91 cell line was found to have the unusual property of being growth-inhibited by insulin (Pawelek *et al.*, 1982). Several variants have been isolated which have survived insulin treatment at 10^{-9} M for growth. From this variant line, variants which no longer require insulin were isolated and these were found to be resistant to insulin. Analysis of cell hybrids made between these lines showed that being insulin-inhibited is dominant over insulin-requiring and insulin-resistant. Molecular analysis remains to be carried out. Similar inhibitory effects of insulin have been observed in human HeLa cells and Chinese hamster lung CHL cells (Shimizu, unpublished).

Growth of human epidermoid carcinoma A431 cells is suppressed by EGF (Barnes, 1982; Buss *et al.*, 1982). Variants that survived at 1×10^{-7} M EGF were isolated. These variants had reduced EGF binding capacity at 8–40% of the capacity of the parental A431 cells. In concert with the decreased binding, EGF-dependent protein kinase activity was reduced. It was suggested from these data that the ability of EGF to kill A431 cells is associated with overwhelming cellular protein phosphorylation. These results also support the proposition made from biochemical studies that the EGF receptor is itself a protein kinase (Cohen *et al.*, 1981, 1982a,b). In these variants, cellular pp60 sarc kinase activity remained unchanged, suggesting that the pp60 sarc kinase is different from the EGF-dependent receptor kinase (Gill and Lazar, 1981).

5.3.4 Use of serum-free media

This selection method relies on the fact that maintenance and growth of cells in serum-free medium require certain combinations of hormones and growth factors. Cell variants that have changed a particular hormone requirement can be isolated on the basis of their ability to grow as colonies in the serum-free hormone-defined media in the absence of that hormone. A completely defined medium that supports good growth of 3T3-L1 cells has been developed (Shimizu *et al.*, 1982; Shimizu, 1983; Gamou and Shimizu, unpublished). The basal nutrient medium is a 3:1 mixture of Dulbecco's modified Eagle's medium and Ham's F12 medium to which is added several amino acids, vitamins, trace elements, bovine crystalline insulin, human transferrin, bovine serum albumin, ethanolamine and EGF. This medium allows one to precisely study the requirements of insulin and EGF for induction and expression of adipocyte differentiation. Similar media have been developed for Swiss/3T3 cells (Shipley and Ham, 1983) and cells with different types of differentiation (Sato *et al.*, 1982). Use of these types of media can aid the isolation of genetic variants and facilitate analysis of the mechanisms involved in differentiation (Shimizu *et al.*, 1982).

5.3.5 Use of hormone– and antibody–toxin conjugates

Bioactive polypeptides can be made toxic without the loss of their receptor-recognizing specificity. This has been done by cross-linking them to the A fragment of diphtheria toxin (DTa) or the A chain of toxic ricin (RICa) (Miskimins and Shimizu, 1979; Shimizu *et al.*, 1980b; Cawley *et al.*, 1980). DTa and RICa are extremely potent inhibitors of protein biosynthesis (Pappenheimer, 1977; Olsnes and Pihl, 1982a,b). After cross-linking to DTa or RICa, the resulting chimaeric polypeptides, such as insulin–DTa or EGF–RICa, have been found to competitively inhibit the binding of the corresponding ligand to its receptors. They were also found to be useful for killing insulin- or EGF-responsive cells in order to specifically select cell variants which are unable to either bind the ligand or take up the bound ligand and process it (Fig. 5.3). The general strategy of the technique for making hybrid proteins is illustrated in Fig. 5.4 using insulin conjugates as examples. Two different cross-linking techniques, both relying upon disulphide linkages, one which

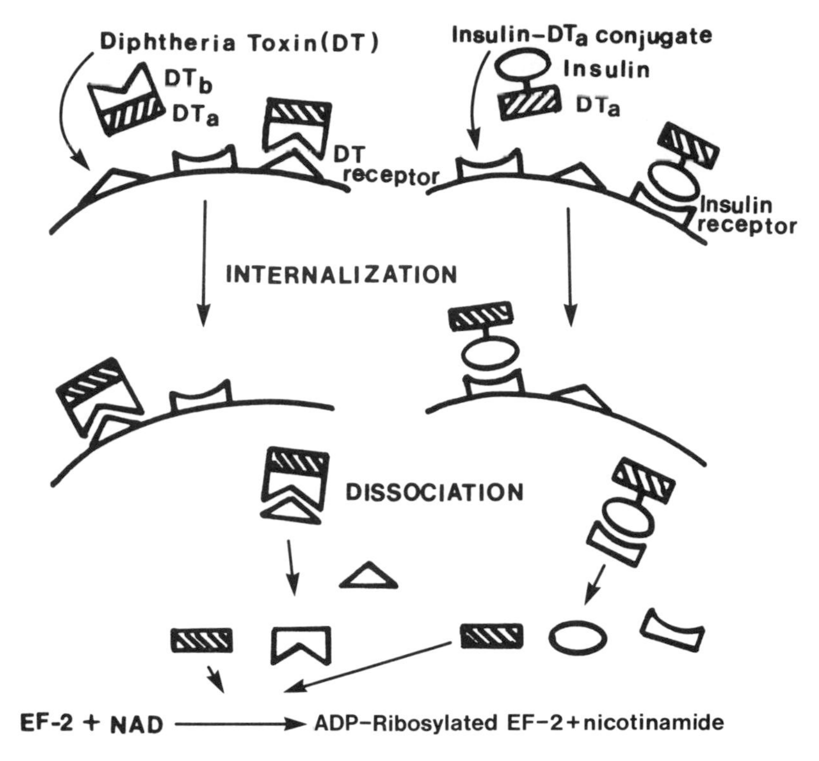

Fig. 5.3 Selection of genetic variants using insulin–toxin A fragment conjugate. The mutants that lack insulin receptors or are deficient in any steps of the presumptive uptake and processing mechanisms will survive, and the normal cells will be killed due to the block of protein biosynthesis. The diphtheria toxin can be split into two fragments. Fragment A is catalytically active, and fragment B is the binding domain for toxin receptor. Insulin was cross-linked via a disulphide bond to fragment A. Processor may be a lysosome. The ADP-ribosylated EF-2 is inactive for chain elongation.

modifies carboxyl groups and one which modifies amino groups of proteins, have been used. Since the amino terminus of insulin is crucial for binding, it is necessary to protect it from the cross-linking reactions. Two different toxin–insulin conjugates have been constructed and used to select several series of resistant variants. Variants of Swiss/3T3 cells which survived the toxicity of insulin–DTa conjugates displayed a deficiency of insulin binding (Miskimins and Shimizu, 1981a). In addition, these variants exhibited alterations in morphology and cell growth properties. These changes were accounted for by insulin's growth regulatory role and the insulin receptor's possible interaction with other membrane components. One of the variants, CI-3, showed an alteration in the lysosomal system or in its regulation (Miskimins *et al.*, 1981). In this case, resistance to the toxic conjugate was correlated with the inefficient processing of the conjugate rather than enhanced degradation in the lysosome despite the accumulation of numerous lysosomes by the CI-3 cell. In fact, no change in the level of lysosomal proteases such as cathepsin D was found (Miskimins and Shimizu, 1981b).

The specificity of the selection against insulin receptors was proved by the finding that the resistant variant lacking insulin binding activity did bind and internalize EGF. DNA synthesis in this variant was not stimulated by insulin but it was initiated by EGF.

The same selection procedure using the insulin–DTa conjugate was applied to two other insulin-responsive cell lines: 3T3-L1 and rat hepatoma H35 cells (Fig. 5.5). The 3T3-L1 cells are able to differentiate into adipocytes under defined induction conditions. Several variants selected from the 3T3-L1 cells

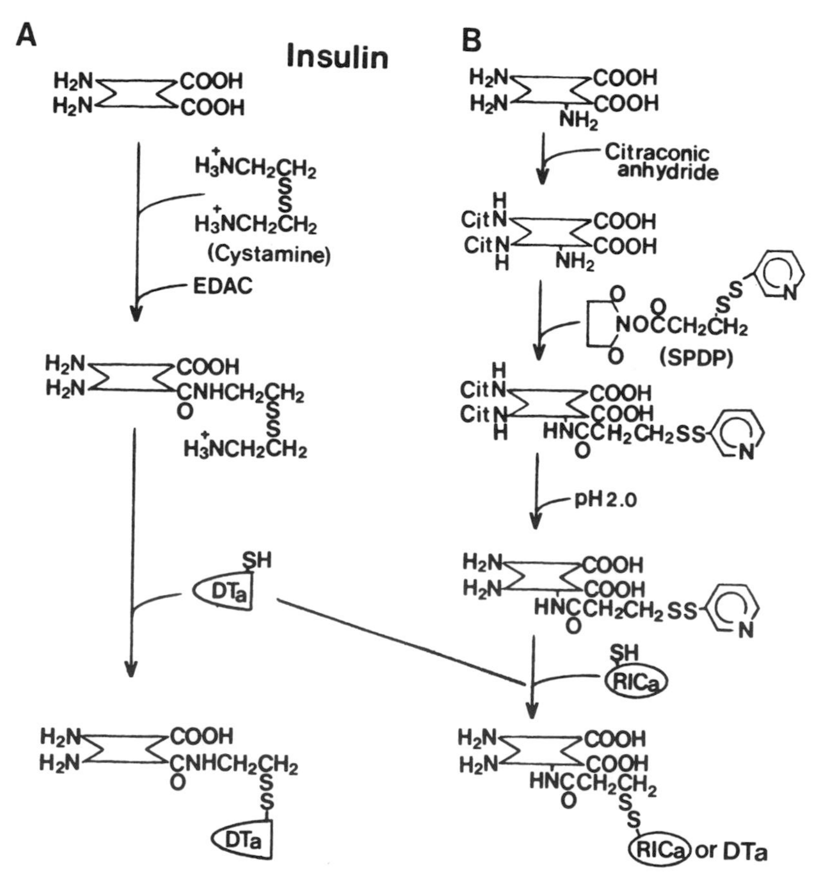

Fig. 5.4 Methods of cross-linking insulin to diphtheria toxin A fragments (DTa) and ricin A chains (RICa). A, the cystamine method to modify carboxyl groups of insulin; B, the SPDP method to modify the amino group of B29 lysine of insulin. See details in the text.

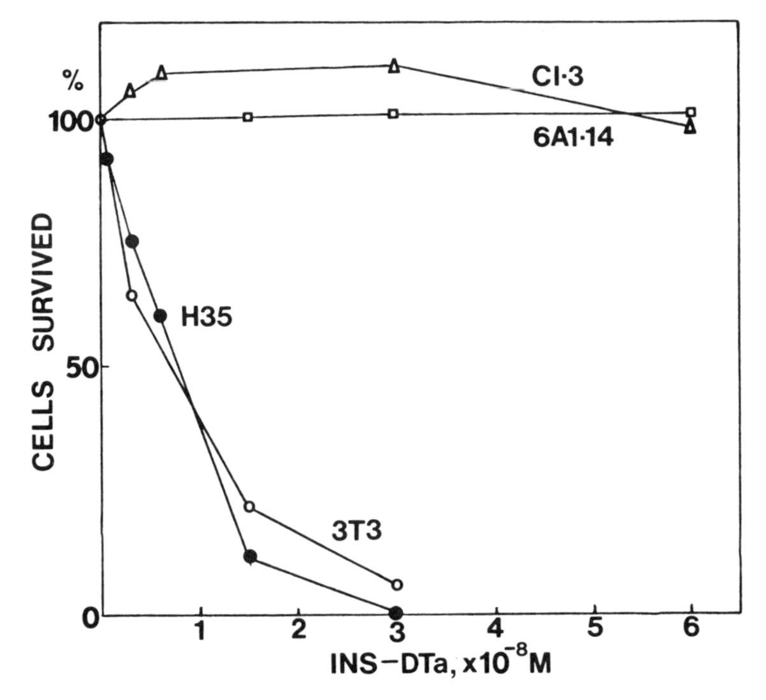

Fig. 5.5 Dose-dependent profiles of cell killing by insulin–DTa conjugates. CI-3 cells were isolated from Swiss/3T3 mouse fibroblasts. 6A 1–14 cells were isolated from H35 rat hepatoma cells. See the text for details.

showed a substantial loss of insulin binding capacity. Interestingly, these variants fell into three categories in terms of their ability to differentiate and in their insulin requirements for growth in serum-free hormone-supplemented medium (Shimizu *et al.*, 1982). The H35 rat hepatoma cells produce tyrosine aminotransferase (TAT) upon exposure to insulin and/or dexamethasone (DEX). Most of the insulin–DTa conjugate-resistant variants of H35 cells again revealed a substantial loss of insulin binding capacity. Some variants apparently have lost insulin-inducible TAT activity and/or insulin-stimulative DNA synthesis (Shimizu, 1983, unpublished). These variants should be able to serve as systems for exploring the genetic control mechanisms of insulin action.

The application of the insulin–RICa conjugate has so far resulted in a few resistant variants (Shimizu, 1983; Gamou and Shimizu, unpublished). These variants showed both reduced and increased levels of insulin binding capacity. Interestingly, one variant, IR-3, was found to be highly resistant to intact ricin.

Protein cross-linking techniques have been used with other polypeptide hormones, such as human placental lactogen, chorionic gonadotropin, epidermal growth factor, and monoclonal antibodies raised against various tumour antigens, EGF receptors and transferrin receptors (Chang and Neville, 1977; Chang *et al.*, 1977; Oeltmann and Heath, 1979; Miskimins and Shimizu, 1979; Cawley *et al.*, 1980; Olsnes and Pihl, 1982a,b). Most, if not all, of these conjugates were proved to be cytotoxic to various cell types but only in a few cases have they been used as an aid for cell variant isolation. When the EGF-RICa conjugate was applied to Chinese hamster lung CHL cells, isolated variants had either elevated or reduced EGF binding capacity. These variants included resistance to intact ricin (Shimizu, Shimizu and Miskimins, unpublished work).

The conjugate of RICa and monoclonal IgG antibody against EGF receptors has provided a unique variant of A431 human carcinoma cells which has apparently lost low affinity EGF receptors, while retaining high affinity EGF receptors (Behzadian and Shimizu, 1983). This variant should be useful for clarifying the EGF receptor/kinase relationship.

These findings clearly indicate that the hormone–toxin conjugates not only allow the selection of variants specifically related to a ligand's activity but also provide variants with a wide spectrum of properties. The reasons for this variety are not completely understood, though discussed in greater detail elsewhere (Shimizu, 1983; Bacha *et al.*, 1983) and await future studies on the mechanism of conjugate entry. Obviously, not only are there many gene products involved in the binding, clustering, internalization and intracellular transport of bioactive polypeptides but also the targeting of the toxic moiety of the conjugates to protein synthesis machinery appears complex. Therefore, a defect in any of these steps might enable the cells to survive in the presence of toxic conjugates. Further application of the hormone–toxin conjugates as selection agents in combination with serum-free hormone-supplemented medium, cell synchronization methods and temporal inhibitor treatment known to block certain steps of the presumptive pathway will allow for the isolation of new types of cell variants.

5.3.6 Use of toxic LDL derivatives

This procedure takes advantage of the unique structure of the low density lipoprotein (LDL) particle, a plasma cholesterol transporter that enters cells by receptor-mediated endocytosis (Goldstein *et al.*, 1979). The LDL particle consists of apoprotein B and a core of cholesteryl ester (Goldstein *et al.*, 1979). The lipid portion can be removed by extraction and the particle can be reconstituted with toxic or fluorescent hydrophobic molecules (Krieger *et al.*, 1981). Because cholesterol is an essential component of mammalian cell membranes, cells cannot survive unless they synthesize cholesterol *de novo* or obtain it exogenously. When cells are grown in the presence of serum they will preferentially utilize the cholesterol available through the culture medium. A cell that has lost the LDL pathway should be resistant to the toxic effects of such agents because of its inability to absorb them from the culture medium. When reconstituted LDL particles containing toxic 25-hydroxycholesteryl oleate (25-HCO) was applied to CHO cells it was taken up via the LDL receptor and converted to the toxic 25-hydroxycholesterol in lysosomes, killing all but the receptor-deficient cells. The resistant cells were further screened with LDL particles reconstituted with a fluorescent cholesteryl ester, r(PMCA oleate) LDL (pyrenemethyl-23,24-dinor-5-cholen-22-oate-3b-yl oleate LDL), and colonies that failed to accumulate fluorescence were isolated. The mutant cells grow in the presence of toxic LDL at concentrations 100 times higher than those that killed the parental cells. Since LDL can be coupled to cell surface binding ligands, this selection method may be generally useful for isolating variants in other receptor systems. Insulin, for example, could be conjugated with LDL apoprotein and reconstituted with toxic 25-HCO, to be used for isolating insulin receptor-deficient cells.

Previously, conditions had been developed in which cell growth was dependent on the LDL pathway. Cells with a mutation in the LDL pathway preventing the uptake of LDL failed to survive under selection conditions and were isolated by negative selection. In order for such a system to operate, it is necessary to completely shut off internal cholesterol synthesis, thus forcing the cells to be dependent upon exogenous sources. This was accomplished through the use of a fungal anti-metabolite, known as compactin, that inhibits HMG-CoA reductase. Appropriate conditions for the use of compactin in cell culture have been determined, but mutants affected in the LDL pathway have not yet been isolated by this method.

When one LDL receptor-deficient mutant clone was co-cultured with other receptor-minus clones, it was induced to express receptors that could mediate normal endocytosis. These LDL receptor-defective clones include two classes of mutations: cbc (complemented by co-cultivation) and icc (inducer cells in co-cultivation). The induction and short-term (18 h) stability of LDL receptors in cbc cells did not require protein synthesis in icc cells. Complementation by co-cultivation only occurred when the cbc and inducing cells were in close proximity, suggesting that an unstable diffusible factor or intimate cell-to-cell association was required for complementation (Krieger, 1983).

5.3.7 Use of lectins and metabolic inhibitors

A number of variants that are resistant to lectins have been isolated (Wright *et al.*, 1980; Stanley, 1980). These include a variety of mutations such as defects in membrane binding sites, intracellular transport ability and alterations in glycosyltransferases and ribosomes. Cell variants resistant to diphtheria toxin (DT) were found to be defective in toxin receptors, elongation factor 2 (EF-2) or diphthamide biosynthesis (Draper *et al.*, 1979; Moehring *et al.*, 1979, 1980). Recent studies suggest that the defects in some of the DT-resistant variants stem from an inability to deliver toxin and lysosomal hydrolases to acidic endocytic vesicles (Robbins *et al.*, 1983; Merion *et al.*, 1983). In one case (Robbins *et al.*, 1983), the defect had minimal effects on uptake and processing of other ligands such as LDL, suggesting a difference in endocytosis pathway for these ligands.

Lysosomal enzymes are synthesized on the rough endoplasmic reticulum (ER), modified, and endocytosed into lysosomes through mannose 6-phosphate receptors (Sly and Fischer, 1982). When CHO cells were treated with ^{125}I-labelled acid phosphatase, the radioactive enzymes were incorporated into wild-type CHO cells which were killed by radioirradiation (Robbins, 1982). The variants that survived this treatment had a deficiency in their lysosomal processing pathway.

Since insulin and EGF receptors are glycoproteins, tunicamycin, an inhibitor of protein glycosylation, inhibits the cell's ability to bind these hormones (Hedo *et al.*, 1981; Stevens *et al.*, 1982). There are a number of metabolic inhibitors that have been determined to affect specific target molecules involved in the receptor–ligand systems. Such drugs as methylamine, monensin and taxol have great potential for the isolation of new kinds of variants.

5.3.8 Use of cell sorting techniques

The fluorescence-activated cell sorter (FACS) has been successfully employed in the separation of cell populations with respect to different surface marker molecules such as differentiation- and tumour-associated markers (Melamed *et al.*, 1979). One particular human–mouse cell hybrid population was separated using a fluorescently labelled antibody raised against a human X-linked surface marker (Dorman *et al.*, 1978). This machine has recently been used to study insulin internalization processes (Murphy *et al.*, 1982) and will be very useful for isolating genetic variants.

5.4 ANALYSIS OF POLYPEPTIDE RECEPTOR SYSTEMS BY CELL HYBRIDIZATION

Somatic cell hybridization has been used to investigate the questions of dominance or recessiveness of cell mutations and of whether complementation between different types of lesions may occur. This approach has been applied to the genetics of glucocorticoid responsiveness (Gehring, 1980; Bourgeois and Pfahl, 1976). However, application of this approach to the genetics of receptor systems for bioactive polypeptides is at present very limited.

Somatic cell genetic analysis of the cell surface receptors is hampered because in most receptor–ligand systems, there is apparently no species specificity in ligand binding. For example, mouse EGF binds to both human and mouse cell receptors. Two methods have been used to overcome this problem: (1) using receptor-deficient mouse cells as the recipient cells in hybridization, and (2) using antibodies which react with the relevant receptor in a species-specific manner. Perhaps the most valuable information so far obtained by cell hybridization analysis is that EGF receptor activity or growth responsiveness is a dominant trait and segregates accordingly in a series of interspecific cell hybrid clones.

5.4.1 Genetic analysis of cellular responsiveness

It has been shown that the ability to respond to the growth-stimulatory action of insulin behaves as a dominant characteristic in cell hybrids (Straus and Williamson, 1978; Coppock *et al.*, 1980). The PG19 mouse melanoma cell line is unresponsive to insulin's growth stimulation in both serum-containing and serum-free conditions. Cell hybrids produced between PG19 cells and mouse embryonic fibroblasts were responsive to insulin's growth effects. Since PG19 had insulin receptors, the recovery of responsiveness in the cell hybrids was attributed to the inheritance from the fibroblast parent. In these cell hybrids stimulation of cell proliferation was correlated with the insulin-dependent phosphorylation of ribosomal S6 protein (Kulkarni and Straus, 1983).

5.4.2 Chromosomal mapping of receptor genes

Fusion of human and mouse cells results in the formation of hybrids that usually retain nearly the full set of mouse chromosomes and a small random subset of human chromosomes (Weiss and Green, 1967). Because both murine and human genes are usually functional, one can assay each hybrid clone for the presence of a given human gene product (Nichols and Ruddle, 1973). For the reasons discussed above, in the study of the EGF receptor, mouse A9 cells deficient in hypoxanthine phosphoribosyltransferase (HPRT) and devoid of EGF binding activity were fused with human diploid fibroblasts possessing EGF binding activity. The human–mouse cell hybrids were isolated after hypoxanthine–aminopterine–thymidine (HAT)–ouabain selection (Shimizu *et al.*, 1980a). Analyses of isozyme markers and chromosomes of a series of hybrid cell clones indicated that EGF binding activity can be restored if human chromosome 7 is retained in the hybrid cells. There was no expression of EGF binding activity without this human chromosome. These initial studies provided the basis to suppose that a gene on human chromosome 7 may be a structural gene for the EGF receptor protein or a gene(s) for modifying the defective mouse receptor protein, if present. A study using immunological approaches clarified the point as to species specificity of the EGF receptors expressed in these cell hybrids. These EGF receptors were in fact of human origin and had a molecular weight of 170 K (Behzadian *et al.*, 1982). Further studies using three series of human–mouse cell hybrid clones which carry unique chromosome rearrangements involving chromosome 7 provided evidence for the regional localization of the EGF receptor gene (EGFR) to the 7p13 → q22 (Kondo and Shimizu, 1983; see Fig. 5.6). It was quite coincidental that antibodies raised against a human cell surface antigen of M_r = 165 K

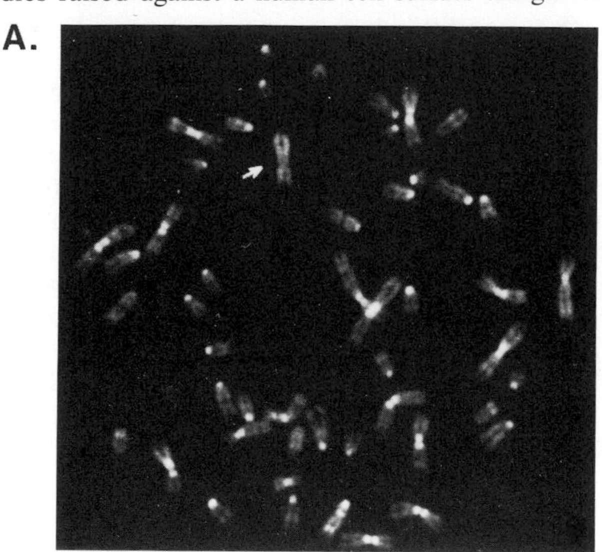

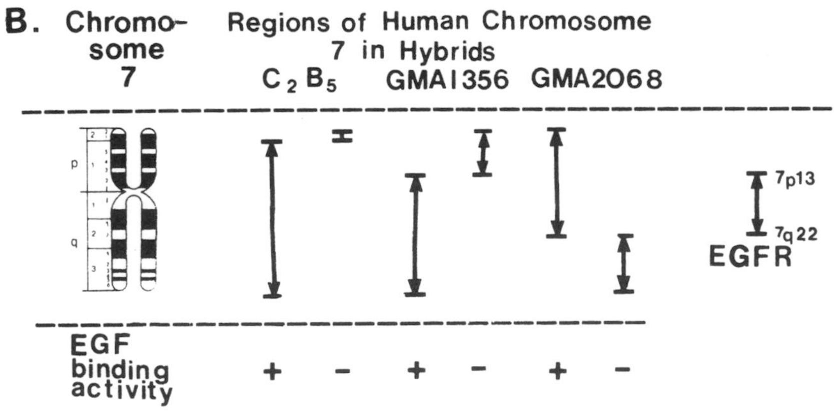

Fig. 5.6 A. Metaphase chromosomes of a human–mouse cell hybrid line C2B5. Chromosomes were stained with a mixed solution of quinacrine mustard and Hoechst 33258. The human X:7 translocation chromosome is indicated by arrow. B. EGF binding activity and the retention of different regions of chromosome 7 in three series of human–mouse cell hybrid lines. The smallest overlapping region can be determined to be 7p13-q22. See the text for details.

(SA7) also immunoprecipitated EGF receptors (Carlin and Knowles, 1982). Since these molecules were proved to be identical and SA7 was known to be localized to the 7pter → p12 (Knowles *et al.*, 1977), it was possible to restrict the gene EGFR to the small region 7p13 → p12 (Kondo and Shimizu, 1983).

The receptor gene for transferrin has recently been mapped to human chromosome 3 (Goodfellow *et al.*, 1982). Transferrin is required for the growth of cells and its effect is particularly noted in serum-free media (Hayashi *et al.*, 1978; Guilbert and Iscove, 1976). This growth-promoting effect of transferrin is probably due to its iron transporting properties. Transferrin is bound specifically to cells by a membrane receptor, a protein of M_r = 180 K which is composed of two disulphide-linked subunits of M_r = 90 K (Seligman *et al.*, 1979; Yang Hu and Aisen, 1978; Fernandez-Pol and Klos, 1980). In this study, both parental cells used for hybrid formation were positive for transferrin binding and therefore all the hybrids expressed binding activity. However, a specific monoclonal antibody (OKT9) that recognizes only the human transferrin receptor has made analysis of the human specific expression pattern possible, thus providing a basis for chromosomal assignment of the transferrin receptor gene (TFR) to chromosome 3 (Goodfellow *et al.*, 1982).

5.4.3 Internalization and processing in cell hybrids

A human–mouse cell hybrid line, C2B5, retains only one human chromosome of an X:7 translocation in addition to the mouse parental genome (Behzadian *et al.*, 1982). The EGF binding profiles and Scatchard analysis indicated that these hybrid cells carry a single class of EGF receptors with high affinity ($K_d = 4\times10^{-10}$ M). These human EGF receptors embedded in the plasma membrane of the hybrid cells, which are essentially composed of mouse components, are subject to 'down regulation' mediated by the ligand EGF. Analysis of the fate of the cell-bound EGF indicted that it is internalized and processed for degradation through a lysosomal pathway. Despite these responses to EGF, DNA synthesis in the C2B5 hybrid cells was barely stimulated (20–30%). When different hybrid clones were tested for a mitogenic response, it was found that retention of human chromosomes other than 7 enabled cells to process EGF more slowly and morphologically become like epithelial cells, gaining stronger mitogenic responsiveness. These results suggest the involvement of multiple chromosomal genes in the 'full' expression of the mitogenic response to EGF.

5.4.4 Genetic analysis of receptor hyperproduction

The human epidermoid carcinoma cell line A431 is unique since the cells possess an extraordinarily high number (2×10^{-6} to 3×10^{6}) of EGF receptors on the plasma membrane (Fabricant *et al.*, 1977). This unique feature has been extremely useful for visualizing the events occurring after binding of EGF to surface receptors, such as clustering of receptors, internalization of the receptor-bound EGF, and its degradative processing through a lysosomal pathway (Haigler *et al.*, 1978). Spontaneous cross-linking of the EGF–receptor complex (Linsley and Fox, 1980), rapid stimulation of fluid pinocytosis by EGF (Haigler *et al.*, 1979), EGF-induced protein phosphorylation in isolated membranes (Carpenter *et al.*, 1978, 1979), dramatic morphological changes of whole cells responding to EGF (Chinkers *et al.*, 1979), and high sensitivity to a cytotoxic EGF cross-linked to diphtheria toxin A fragment (Shimizu *et al.*, 1980b) have also been demonstrated using A431 cells. Despite the extensive use of A431 cells for cell biological analysis, little was known about their genetic constitution. Recently, chromosome analysis revealed that A431 cells are hypotetraploid and possess two copies of intact chromosome 7 and two types of the translocation chromosomes involving chromosome 7 (M4 and M14) as well as several other chromosome rearrangements. The A431 cells were fused with mouse A9 cells, which lack EGF receptors and are deficient in HPRT, and the A431–A9 cell hybrids were selected in HAT–ouabain medium (Shimizu *et al.*, 1983c). In these AA-series hybrid clones it was found that the expression of high EGF binding activity was correlated with the presence of human translocation chromosome M4. AA-hybrid clones that contained intact human chromosome 7 but not the marker chromosome M4 expressed only ordinary levels of EGF receptors. The EGF receptors expressed in AA-hybrids were proved to be of human nature by immunoprecipitation. From these observations of the ^{125}I-EGF–receptor complex it was suggested that the marker chromosome M4 may carry an alteration(s) in the gene(s) involved in EGF receptor biosynthesis. These AA-hybrids have provided a unique system for studying the genetic regulation of EGF receptor gene expression. Complete understanding at molecular levels, however, awaits further investigation.

The EGF receptor has been found to be tightly associated with a protein kinase activity and it has been postulated that the EGF receptor molecule contains a domain for a tyrosine-specific protein kinase (Cohen *et al.*, 1980). A correlation was found in several variants of A431 cells between the reduction of EGF receptor kinase activity and the loss of the marker chromosome M4 (Gill *et al.*, 1983).

5.4.5 Other uses of variant cells

The EGF receptor-negative variants of 3T3 cells have made possible the intriguing finding that an exogenous EGF receptor can be inserted into a recipient cell (Bishayee *et al.*, 1982). When NR-6 cells were incubated with EGF receptor-rich mouse hepatic membranes, there was a transfer of almost 20% of the input of EGF receptors to the recipient cells, whereas only 1–2% of bulk hepatic proteins were transferred. The inserted receptor bound EGF with high affinity, and the bound EGF stimulated DNA synthesis and cell division.

The uptake of EGF receptors was proved not to be due to an activation of endogenous protein synthesis in NR-6 cells. The existence of a cellular mechanism for specific receptor transfer was proposed.

5.5 SUMMARY AND PERSPECTIVES

In this chapter, I have reviewed the current status of our knowledge of the cellular and molecular aspects of bioactive polypeptide action. Discussion of the growth-stimulatory effects of insulin was particularly focused on comparisons with the effects of EGF and other cell growth factors. Complete understanding of their molecular action awaits comprehensive studies from different points of view. I have discussed the emerging cell genetic approach toward understanding the genetic basis of bioactive polypeptide–receptor systems. It became evident from these studies that not only the loss of ligand binding activity but also defects in post-receptor processing inevitably result in the loss of cellular responsiveness. The use of genetic variants isolated by a number of different selection methods in biochemical and cell hybridization analyses has provided valuable information, particularly about the chromosomal localization of the receptor genes. These variants can also be used for studying genetic control of receptor–ligand systems. Future studies will clearly depend on more intensive use of already established selection methods as well as the development of new procedures for the selection of variants.

Recombinant DNA technology (Abelson, 1980) has been used for studying the structural organization of several polypeptide hormone genes such as insulin (Ullrich *et al.*, 1977). It will be of great importance to apply these techniques to the study of receptor genes as well. It would be exciting to attempt transfer of particular genes into a variety of variants (Wigler *et al.*, 1979) to analyse their effects on the mitogenic responsiveness. Further use of anti-receptor monoclonal antibodies will aid in the analysis of structural alterations in the receptors of cell variants. The recent discovery of a family of growth factors and transforming growth factors (Roberts *et al.*, 1983) implies a complex genetic regulation of their receptor systems. The cell genetic approach will explore these problems in addition to receptor biosynthesis, mechanisms of subunit assembly and receptor insertion into the plasma membrane.

Receptor-deficiency diseases have been discovered. In the future, cell genetic studies will facilitate the diagnosis of deficiencies in polypeptide–receptor systems and may uncover new types of human genetic disorders that are related to post-receptor mechanisms. This in turn may aid in developing new ways to treat or even cure these disorders.

REFERENCES

Abelson, J. (1980), *Science*, **209**, 1319–1321.

Adamson, E.D. and Rees, A.R. (1981), *Mol. Cell. Biochem.*, **34**, 129–152.

Allegra, J.C. and Lippman, M.E. (1978), *Cancer Res.*, **38**, 3823–3829.

Anderson, R.G., Brown, M.S. and Goldstein, J.L. (1977), *Cell*, **10**, 351–364.

Bacha, P., Murphy, J.R. and Reichlin, S. (1983), *J. Biol. Chem.*, **258**, 1565–1570.

Baker, J.B., Simmer, R.L., Glenn, K.C. and Cunningham, D.D. (1979), *Nature (London)*, **278**, 743–745.

Barnes, D.W. (1982), *J. Cell Biol.*, **93**, 1–4.

Barnes, D. and Sato, G. (1979), *Nature (London)*, **281**, 388–389.

Barnes, D. and Sato, G. (1980), *Cell*, **22**, 649–655.

Behzadian, M.A. and Shimizu, N. (1983), *J. Cell Biol.*, **97**, 408a.

Behzadian, M.A., Shimizu, Y., Kondo, I. and Shimizu, N. (1982), *Somat. Cell Genet.*, **8**, 347–362.

Bergeron, J.J.M., Sikstrom, R., Hand, A.R. and Posner, B.I. (1979), *J. Cell Biol.*, **80**, 427–443.

Berhanu, P., Olefsky, J.M., Tsai, P., Thamm, P., Saunders, D. and Brandenburg, D. (1982), *Proc. Natl. Acad. Sci. U.S.A.*, **79**, 4069–4073.

Bishayee, S., Feinman, J., Michael, H., Pittenger, M. and Das, M. (1982), *Proc. Natl. Acad. Sci. U.S.A.*, **79**, 1893–1897.

Blundell, T.L., Bedarkar, S. and Humbel, R.E. (1983), *Fed. Proc.*, **42**, 2592–2597.

Bourgeois, S. and Pfahl, M. (1976), *Adv. Protein Chem.*, **30**, 1–99.

Bradley, W.E. and Culp, L.A. (1974), *Exp. Cell Res.*, **84**, 335–350.

Brown, K.D., Dicker, P. and Rozengurt, E. (1979), *Biochem. Biophys. Res. Commun.*, **86**, 1037–1043.

Brown, M.S., Anderson, R.G.W. and Goldstein, J.L. (1983), *Cell*, **32**, 663–667.

Bulter-Gralla, E. and Herschman, H.R. (1981), *J. Cell Physiol.*, **107**, 59–67.

Buss, J.E., Kudlow, J.E., Lazar, C.S. and Gill, G.N. (1982), *Proc. Natl. Acad. Sci. U.S.A.*, **79**, 2574–2578.

Bylund, D.B. (1980), in *Receptor Binding Techniques*, American Society for Neuroscience, pp. 70–99.

Carlin, C.R. and Knowles, B.B. (1982), *Proc. Natl. Acad. Sci. U.S.A.*, **79**, 5026–5030.

Carpenter, G. and Cohen, S. (1976), *J. Cell Biol.*, **71**, 159–171.

Carpenter, G., Lembach, K.J., Morrison, M.M. and Cohen, S. (1975), *J. Biol. Chem.*, **250**, 4297–4304.

Carpenter, G., King, L., Jr. and Cohen, S. (1978), *Nature (London)*, **276**, 409–410.

Carpenter, G., King, L., Jr. and Cohen, S. (1979), *J. Biol. Chem.*, **254**, 4884–4891.

Castagna, M., Takai, Y., Kaibuchi, K., Sano, K., Kikkawa, U. and Nishizuka, Y. (1982), *J. Biol. Chem.*, **257**, 7847–7851.

Cawley, D.B., Herschman, H.R., Gilliland, D.G. and Collier, R.J. (1980), *Cell*, **22**, 563–570.

Chafouleas, J.G., Bolton, W.E., Hidaka, H., Boyd, A.E. and Means, A.R. (1982), *Cell*, **28**, 41–50.

Chambard, J.-C., Franchi, A., Le Cam, A. and Pouyssegur, J. (1983), *J. Biol. Chem.*, **258**, 1705–1713.

Chang, T.-M. and Neville, D.M., Jr. (1977), *J. Biol. Chem.*, **252**, 1505–1514.

Chang, T.-M., Dazord, A. and Neville, D.M., Jr. (1977), *J. Biol. Chem.*, **252,** 1515–1522.

Cheung, W.K. (1980), *Science*, **207,** 19–27.

Chinkers, M. and Cohen, S. (1981), *Nature (London)*, **290,** 516–519.

Chinkers, M., McKanna, J.A. and Cohen, S. (1979), *J. Cell Biol.*, **83,** 260–265.

Cohen, P. (1982), *Nature (London)*, **296,** 613–620.

Cohen, S., Carpenter, G. and King, L., Jr. (1980), *J. Biol. Chem.*, **255,** 4834–4842.

Cohen, S., Chinkers, M. and Ushiro, H. (1981), in *Protein Phosphorylation* (O.M. Rosen and E.G. Krebs, eds), Cold Spring Harbor Laboratory, New York, pp. 801–808.

Cohen, S., Fava, R.A. and Sawyer, S.T. (1982a), *Proc. Natl. Acad. Sci. U.S.A.*, **79,** 6237–6241.

Cohen, S., Ushiro, H., Stoscheck, C. and Chinkers, M. (1982b), *J. Biol. Chem.*, **257,** 1523–1531.

Colburn, N.H., Wendel, E.J. and Abruzzo, G. (1981), *Proc. Natl. Acad. Sci. U.S.A.*, **78,** 6912–6916.

Cooper, G.M. (1982), *Science*, **218,** 801–806.

Cooper, J.A. and Hunter, T. (1981), *J. Cell Biol.*, **91,** 878–883.

Coppock, D.L., Covey, L.R. and Straus, D.S. (1980), *J. Cell Physiol.*, **105,** 81–92.

Crossin, K.L. and Carney, D.H. (1981), *J. Cell Biol.*, **91,** 334a.

Cushman, S.W. and Wardzala, L.J. (1980), *J. Biol. Chem.*, **255,** 4758–4762.

Czech, M.P., Oppenheimer, C.L. and Massague, J. (1983), *Fed. Proc.*, **42,** 2598–2601.

Das, M. and Fox, F. (1978), *Proc. Natl. Acad. Sci. U.S.A.*, **75,** 2644–2648.

Das, M., Miyakawa, T., Fox, C.F., Pruss, R.M., Aharonov, A. and Herschman, H.R. (1977), *Proc. Natl. Acad. Sci. U.S.A.*, **74,** 2790–2794.

Davies, R.L., Grosse, V.A., Kucherlapati, R. and Bothwell, M. (1980), *Proc. Natl. Acad. Sci. U.S.A.*, **77,** 4188–4192.

DeAsua, L.J., O'Farrel, M.K., Clingan, D. and Rudland, P.S. (1977a), *Proc. Natl. Acad. Sci. U.S.A.*, **74,** 3845–3849.

DeAsua, L.J., O'Farrel, M.K., Bennett, D., Clingan, D. and Rudland, P.S. (1977b), *Nature (London)*, **265,** 151–153.

DeMeyts, P., Roth, J., Neville, D.M., Jr., Gavin, J.R., III and Lesniak, M.A. (1973), *Biochem. Biophys. Res. Commun.*, **55,** 154–161.

Dickson, R.B., Nicolas, J.-C., Willingham, M.C. and Pastan, I. (1981), *Exp. Cell Res.*, **132,** 488–493.

Dorman, B.P., Shimizu, N. and Ruddle, F.H. (1978), *Proc. Natl. Acad. Sci. U.S.A.*, **75,** 2363–2367.

Draper, R.K., Chin, D., Eurey-Owens, D., Scheffler, I.E. and Simon, M.I. (1979), *J. Cell Biol.*, **83,** 116–125.

Fabricant, R.N., DeLarco, J.E. and Todaro, G.J. (1977), *Proc. Natl. Acad. Sci. U.S.A.*, **74,** 565–569.

Fain, J.N. (1980) in *Biochemical Actions of Hormones*, Vol. VII, Academic Press, London, pp. 119–203.

Fehlmann, M., Carpentier, J.L., LeCam, A., Thamm, P., Saunders, D., Brandenburg, D., Orci, L. and Freychet, P. (1982), *J. Cell Biol.*, **93,** 82–87.

Fehlmann, M., Carpentier, J.L., Van Obberghen, E., Freychet, P., Thamm, P., Saunders, D., Brandenburg, D. and Orci, L. (1982), *Proc. Natl. Acad. Sci. U.S.A.*, **79,** 5921–5925.

Fernandez-Pol, J.A. and Klos, D.J. (1980), *Biochemistry,* **19,** 3904–3912.
Ferro, A.M., Higgins, N.P. and Olivera, B.M. (1983), *J. Biol. Chem.,* **258,** 6000–6003.
Fine, R.E., Goldenberg, R., Sorrentino, J. and Herschman, H.A. (1981), *J. Supramol. Struct.,* **10,** 199–214.
Fujiki, H., Mori, M., Nakayasu, M., Terada, M., Sugimura, T. and Moore, R.E. (1981), *Proc. Natl. Acad. Sci. U.S.A.,* **78,** 3872–3876.
Gates, R.E. and King, L., Jr. (1982), *Mol. Cell Endocrinol.,* **27,** 263–276.
Gehring, U. (1980), in *Biochemical Actions of Hormones,* Vol. VII, Academic Press, London, pp. 205–232.
Gey, G.O. and Thalhimer, W. (1924), *J. Am. Med. Assoc.,* **82,** 1609–1612.
Gill, G.N. and Lazar, C.S. (1981), *Nature (London),* **293,** 305–307.
Gill, G.N., Lazar, C.S., Gamo, S. and Shimizu, N. (1983), *American Society for Clinical Investigation,* Abstracts.
Glass, D.B. and Krebs, E.G. (1980), *Annu. Rev. Pharmacol. Toxicol.,* **20,** 363–388.
Glenn, K., Bowen-Pope, D.F. and Ross, R. (1982), *J. Biol. Chem.,* **257,** 5172–5176.
Goewert, R.R., Klaven, N.B. and McDonald, J.M. (1983), *J. Biol. Chem.,* **258,** 9995–9999.
Goldfine, I.D. (1981a), *Biochim. Biophys. Acta,* **650,** 53–67.
Goldfine, I.D. (1981b), in *Biochemical Actions of Hormones,* Vol. VIII, Academic Press, London, pp. 273–305.
Goldfine, I.D., Smith, G.J., Wong, K.Y. and Jones, A.L. (1977), *Proc. Natl. Acad. Sci. U.S.A.,* **74,** 1368–1372.
Goldstein, J.L., Anderson, R.G.W. and Brown, M.S. (1979), *Nature (London),* **279,** 679–685.
Goodfellow, P.N., Banting, G., Sutherland, R., Greaves, M., Solomon, E. and Povey, S. (1982), *Somat. Cell Genet.,* **8,** 197–206.
Gorden, P., Carpentier, J.-L., LeCam, A., Freychet, P. and Orci, L. (1978), *Science,* **200,** 782–785.
Gorden, P., Carpentier, J.-L., Fan, J.-Y. and Orci, L. (1982a), *Metabolism,* **31,** 664–669.
Gorden, P., Carpentier, J.-L., Moule, M.L., Yip, C.C. and Orci, L. (1982b), *Diabetes,* **31,** 659–662.
Guilbert, L.J. and Iscove, N.N. (1976), *Nature (London),* **263,** 594–595.
Haigler, H., Ash, J.F., Singer, S.J. and Cohen, S. (1978), *Proc. Natl. Acad. Sci. U.S.A.,* **75,** 3317–3321.
Haigler, H.T., McKanna, J.A. and Cohen, S. (1979), *J. Cell Biol.,* **81,** 382–395.
Haigler, H.T., Maxfield, F.R., Willingham, M.C. and Pastan, I. (1980a), *J. Biol. Chem.,* **255,** 1239–1241.
Haigler, H.T., Willingham, M.C. and Pastan, I. (1980b), *Biochem. Biophys. Res. Commun.,* **94,** 630–637.
Hayashi, I., Larner, J. and Sato, G. (1978), *In Vitro,* **14,** 23–30.
Hedo, J.A., Kasuga, M., Van Obberghen, E., Roth, J. and Kahn, C.R. (1981), *Proc. Natl. Acad. Sci. U.S.A.,* **78,** 4791–4795.
Heldin, C.-H., Ek, B. and Rönnstrand, L. (1983), *J. Biol. Chem.,* **258,** 10054–10061.
Hock, R.A., Nex, E. and Hollenberg, M.D. (1979), *Nature (London),* **277,** 403–405.
Hunter, T. and Cooper, J.A. (1981), *Cell,* **24,** 741–752.
Jacobs, S. and Cuatrecasas, P. (1981), *Endocrinol. Rev.,* **2,** 251–263.

Jacobs, S., Hazum, E., Shechter, Y. and Cuatrecasas, P. (1979), *Proc. Natl. Acad. Sci. U.S.A.*, **76**, 4918–4921.

Jacobs, S., Kull, F.C., Jr., Earp, H.S., Svoboda, M.E., VanWyk, J.J. and Cuatrecasas, P. (1983), *J. Biol. Chem.*, **258**, 9581–9584.

Jarett, L. and Seals, J.R. (1979), *Science*, **206**, 1407–1408.

Jarett, L., Kiechle, F.L., Popp, D.A. and Kotagal, N. (1981), in *Protein Phosphorylation* (O.M. Rosen and E.G. Krebs, eds), Cold Spring Harbor Laboratory, pp. 715–726.

Johnson, L.K., Vlodavsky, I., Baxter, J.D. and Gospodarowicz, D. (1980), *Nature (London)*, **287**, 340–343.

Kaneko, Y. (1983), *Biochim. Biophys. Acta*, **762**, 111–118.

Kasuga, M., Van Obberghen, E., Nissley, S.P. and Rechler, M.M. (1981), *J. Biol. Chem.*, **256**, 5305–5308.

Kasuga, M., Karlsson, F.A. and Kahn, C.R. (1982a), *Science*, **215**, 185–186.

Kasuga, M., Van Obberghen, E., Nisley, S.P. and Rechler, M.M. (1982b), *Proc. Natl. Acad. Sci. U.S.A.*, **79**, 1864–1868.

Kasuga, M., Zick, Y., Blithe, D.L., Crettaz, M. and Kahn, C.R. (1982c), *Nature (London)*, **298**, 667–669.

Khan, M.N., Posner, B.I., Verma, A.K., Khan, R.J. and Bergeron, J.J.M. (1981), *Proc. Natl. Acad. Sci. U.S.A.*, **78**, 4980–4984.

Khan, M.N., Posner, B.I., Khan, R.J. and Bergeron, J.J.M. (1982), *J. Biol. Chem.*, **257**, 5969–5976.

King, A.C. and Cuatrecasas, P. (1982), *J. Biol. Chem.*, **257**, 3053–3060.

King, A.C., Hernaez-Davis, L. and Cuatrecasas, P. (1980a), *Proc. Natl. Acad. Sci. U.S.A.*, **77**, 3283–3287.

King, G.L., Kahn, C.R., Rechler, M.M. and Nissley, S.P. (1980b), *J. Clin. Invest.*, **66**, 130–140.

Knowles, B.B., Solter, D., Trinchieri, G., Maloney, K.M., Ford, S.R. and Aden, D.P. (1977), *J. Exp. Med.*, **145**, 314–326.

Kondo, I. and Shimizu, N. (1983), *Cytogenet. Cell Genet.*, **35**, 9–14.

Kono, T., Robinson, F.W., Blevins, T.L. and Ezaki, O. (1982), *J. Biol. Chem.*, **257**, 10942–10947.

Koontz, J.W. and Iwahashi, M. (1981), *Science*, **211**, 947–949.

Krieger, M. (1983), *Cell*, **33**, 413–422.

Krieger, M., Brown, M.S. and Goldstein, J.L. (1981), *J. Mol. Biol.*, **150**, 167–184.

Kudlow, J.E., Buss, J.E. and Gill, G.N. (1981), *Nature (London)*, **290**, 519–521.

Kulkarni, R.K. and Straus, D.S. (1983), *Biochim. Biophys. Acta*, **762**, 542–551.

Larner, J., Cheng, K., Huang, L. and Galasko, G. (1981), in *Protein Phosphorylation* (O.M. Rosen and E.G. Krebs, eds), Cold Spring Harbor Laboratory, New York, pp. 727–733.

Lee, L.-S. and Weinstein, I.B. (1978), *Science*, **202**, 313–315.

Lefkowitz, R.J., Sharp, G.W.G. and Haber, E. (1973), *J. Biol. Chem.*, **248**, 342–349.

Linsley, P.S. and Fox, C.F. (1980), *J. Supramol. Struct.*, **14**, 441–459.

Magun, B.E., Matrisian, L.M. and Bowden, G.T. (1980), *J. Biol. Chem.*, **255**, 6373–6381.

Maness, P.F. and Walsh, R.C. (1982), *Cell*, **30**, 253–262.

Marshall, S. and Olefsky, J.M. (1979), *J. Biol. Chem.*, **254**, 10153–10160.

Marshall, S., Green, A. and Olefsky, J.M. (1981), *J. Biol. Chem.*, **256**, 11464–11470.
Massague, J. and Czech, M.P. (1982), *J. Biol. Chem.*, **257**, 5038–5045.
Massague, J., Pilch, P.F. and Czech, M.P. (1980), *Proc. Natl. Acad. Sci. U.S.A.*, **77**, 7137–7141.
Massague, J., Guillette, B.J. and Czech, M.P. (1981a), *J. Biol. Chem.*, **256**, 2122–2125.
Massague, J., Pilch, P.F. and Czech, M.P. (1981b), *J. Biol. Chem.*, **256**, 3182–3190.
Massague, J., Blinderman, L.A. and Czech, M.P. (1982a), *J. Biol. Chem.*, **257**, 13958–13963.
Massague, J., Czech, M.P., Iwata, K., DeLarco, J.E. and Todaro, G.J. (1982b), *Proc. Natl. Acad. Sci. U.S.A.*, **79**, 6822–6826.
Mastro, A.M. (1982), in *Lymphokines*, Academic Press, London, pp. 263–313.
McKanna, J.A., Haigler, H.T. and Cohen, S. (1979), *Proc. Natl. Acad. Sci. U.S.A.*, **76**, 5689–5693.
McKeehan, W.L., McKeehan, K.A. and Calkins, D. (1982), *Exp. Cell Res.*, **140**, 25–30.
Means, A.R. and Dedman, J.R. (1980), *Nature (London)*, **285**, 73–77.
Melamed, M.R., Mullaney, P.F. and Mendelsohn, M.L. (1979), *Flow Cytometry and Sorting*, J. Wiley & Sons, New York.
Merion, M., Schlesinger, P., Brooks, R.M., Moehring, J.M., Moehring, T.J. and Sly, W.S. (1983), *Proc. Natl. Acad. Sci. U.S.A.*, **80**, 5315–5319.
Miskimins, W.K. and Shimizu, N. (1979), *Biochem. Biophys. Res. Commun.*, **91**, 143–151.
Miskimins, W.K. and Shimizu, N. (1981a), *Proc. Natl. Acad. Sci., U.S.A.*, **78**, 445–449.
Miskimins, W.K. and Shimizu, N. (1981b), *J. Cell Biol.*, **91**, Abstr. no. 2, part 2, p. 216a.
Miskimins, W.K. and Shimizu, N. (1982a), *J. Cell Biochem.*, **20**, 41–50.
Miskimins, W.K. and Shimizu, N. (1982b), *J. Cell Physiol.*, **112**, 327–338.
Miskimins, W.K. and Shimizu, N. (1983), *Proc. 8th Am. Peptide Symp.*, in press.
Miskimins, W.K., Ferris, W.R. and Shimizu, N. (1981), *J. Supramol. Struct. Cell. Biochem.*, **16**, 105–113.
Miskimins, R., Miskimins, W.K., Bernstein, H. and Shimizu, N. (1983), *Exp. Cell Res.*, **146**, 53–62.
Moehring, T.J., Danley, D.E. and Moehring, J.M. (1979), *Somat. Cell Genet.*, **5**, 469–480.
Moehring, J.M., Moehring, T.J. and Danley, D.E. (1980), *Proc. Natl. Acad. Sci. U.S.A.*, **77**, 1010–1014.
Murphy, R.F., Powers, S., Verderame, M., Cantor, C.R. and Pollack, R. (1982), *Cytometry*, **2**, 402–406.
Nagarajan, L., Nissley, S.P., Rechler, M.M. and Anderson, W.B. (1982), *Endocrinology*, **110**, 1231–1237.
Nichols, E.A. and Ruddle, F.H. (1973), *J. Histochem. Cytochem.*, **21**, 1066–1081.
Nilsen-Hamilton, M., Hamilton, R.T., Allen, W.R. and Potter-Perigo, S. (1982), *Cell*, **31**, 237–242.
Nishizuka, Y. and Takai, Y. (1981), in *Protein Phosphorylation* (O.M. Rosen and E.G. Krebs, eds), Cold Spring Harbor Laboratory, New York, pp. 237–249.
Oeltmann, T.N. and Heath, E.C. (1979), *J. Biol. Chem.*, **254**, 1028–1032.
Olsnes, S. and Pihl, A. (1982a), in *Molecular Action of Toxins and Viruses* (S. Van Heyningen and P. Cohen, eds), Elsevier/North Holland, Amsterdam, pp. 51–105.

Olsnes, S. and Pihl, A. (1982b), *Pharmacol. Ther.*, **15,** 355–381.

Oppenheimer, C.L., Pessin, J.E., Massague, J., Gitomer, W. and Czech, M.P. (1983), *J. Biol. Chem.*, **258,** 4824–4830.

Osborne, C.K., Monaco, M.E., Lippman, M.E. and Kahn, C.R. (1978), *Cancer Res.*, **38,** 94–102.

Otto, A.M. (1982), *Cell Biol. Int. Rep.*, **6,** 1–18.

Pappenheimer, A.M., Jr. (1977), *Annu. Rev. Biochem.*, **46,** 69–94.

Pastan, I.H. and Willingham, M.C. (1981a), *Annu. Rev. Physiol.*, **43,** 239–250.

Pastan, I.H. and Willingham, M.C. (1981b), *Science,* **214,** 504–509.

Pawelek, J., Murray, M. and Fleischmann, R. (1982), in *Growth of Cells in Hormonally Defined Media* (G.H. Sato, A.B. Pardee and D.A. Sirbasku, eds), Cold Spring Harbor Laboratory, New York, pp. 911–919.

Pike, L.J., Gallis, B., Casnellie, J.E., Bornstein, P. and Krebs, E.G. (1982), *Proc. Natl. Acad. Sci. U.S.A.*, **79,** 1443–1447.

Pilch, P.F. and Czech, M.P. (1980), *J. Biol. Chem.*, **256,** 1722–1731.

Posner, B.I., Raquidan, D., Josefsberg, Z. and Bergeron, J.J.M. (1978), *Proc. Natl. Acad. Sci. U.S.A.*, **75,** 3302–3306.

Posner, B.I., Patel, B.A., Khan, M.N. and Bergeron, J.J.M. (1982), *J. Biol. Chem.*, **257,** 5789–5799.

Pruss, R.M. and Herschman, H.R. (1977), *Proc. Natl. Acad. Sci. U.S.A.*, **74,** 3918–3921.

Purrello, F., Burnham, D.B. and Goldfine, I.D. (1983), *Proc. Natl. Acad. Sci. U.S.A.*, **80,** 1189–1193.

Rechler, M.M., Schilling, E.E., King, G.S., Maoili, F., Rosenberg, A.M., Higa, O.Z., Podskalny, J.M., Grunfeld, C., Nissley, S.P. and Kahn, C.R. (1980), in *Receptors for Neurotransmitters and Peptide Hormones* (G. Pepen, J. Kuhar and S.J. Enna, eds), Raven Press, New York, pp. 489–497.

Robbins, A.R. (1982), *J. Cell. Biochem. Suppl.*, **6,** 120.

Robbins, A.R., Peng, S.S. and Marshall, J.L. (1983), *J. Cell Biol.*, **96,** 1064–1071.

Roberts, A.B., Frolik, C.A., Anzano, M.A. and Sporn, M.B. (1983), *Fed. Proc.*, **42,** 2621–2626.

Rodbell, M., Krans, H.M.J., Phol, S.L. and Birnbaumer, L. (1971), *J. Biol. Chem.*, **246,** 1872–1876.

Ronnett, G.V., Knutson, V.P. and Lane, M.D. (1982), *J. Biol. Chem.*, **257,** 4285–4291.

Ronnett, G.V., Tennekoon, G., Knutson, V.P. and Lane, M.D. (1983), *J. Biol. Chem.*, **258,** 283–290.

Roth, J. and Grunfeld, C. (1981), in *Williams Textbook of Endocrinology* (R.H. Williams, ed.), Saunders, Philadelphia, pp. 15–72.

Roth, J., Lesniak, M.A., Megyesi, K. and Kahn, C.R. (1979), in *Hormones and Cell Culture* (G.H. Sato and R. Ross, eds), Cold Spring Harbor Laboratory, New York, pp. 167–186.

Rozengurt, E. (1980), in *Curr. Top. Cell. Regul.*, **17,** 59–88.

Saltiel, A.R., Siegel, M.I., Jacobs, S. and Cuatrecasas, P. (1982), *Proc. Natl. Acad. Sci. U.S.A.*, **79,** 3513–3517.

Saltiel, A.R., Doble, A., Jacobs, S. and Cuatrecasas, P. (1983), *Biochem. Biophys. Res. Commun.*, **110,** 789–795.

Sano, K., Takai, Y., Yamanishi, J. and Nishizuka, Y. (1983), *J. Biol. Chem.*, **258,** 2010–2013.

Sato, G. and Ross, R. (1979), *Hormones and Cell Culture*, Cold Spring Harbor Laboratory, New York.

Sato, G.H., Pardee, A.B. and Sirbasku, D.A. (1982), *Growth of Cells in Hormonally Defined Media,* Cold Spring Harbor Laboratory, New York.

Schechter, Y., Hernaez, L. and Cuatrecasas, P. (1978), *Proc. Natl. Acad. Sci. U.S.A.*, **75,** 5788–5791.

Schiff, P.B., Tiant, J. and Horwitz, S.B. (1979), *Nature (London)*, **277,** 665–667.

Schlessinger, J., Schechter, Y., Willingham, M.C. and Pastan, I. (1978a), *Proc. Natl. Acad. Sci. U.S.A.*, **75,** 2659–2663.

Schlessinger, J., Schechter, Y., Cuatrecasas, P., Willingham, M.C. and Pastan, I. (1978b), *Proc. Natl. Acad. Sci. U.S.A.*, **75,** 5353–5357.

Schreiber, A.B., Lax, I., Yarden, Y., Eshhar, Z. and Schlessinger, J. (1981a), *Proc. Natl. Acad. Sci. U.S.A.*, **78,** 7535–7539.

Schreiber, A.B., Yarden, Y. and Schlessinger, J. (1981b), *Biochem. Biophys. Res. Commun.*, **101,** 517–523.

Schreiber, A.B., Libermann, T.A., Lax, I., Yarden, Y. and Schlessinger, J. (1983). *J. Biol. Chem.*, **258,** 846–853.

Seals, J.R. and Czech, M.P. (1980), *J. Biol. Chem.*, **255,** 6529–6531.

Seals, J.R., McDonald, J.M. and Jarett, L. (1979a), *J. Biol. Chem.*, **254,** 6991–6996.

Seals, J.R., McDonald, J.M. and Jarett, L. (1979b), *J. Biol. Chem.*, **254,** 6997–7001.

Seligman, P.A., Schleicher, R.B. and Allen, R.H. (1979), *J. Biol. Chem.*, **254,** 9943–9946.

Shimizu, N. (1983), *Method Cell. Mol. Biol.*, in press.

Shimizu, N., Behzadian, M.A. and Shimizu, Y. (1980a), *Proc. Natl. Acad. Sci. U.S.A.*, **77,** 3600–3604.

Shimizu, N., Miskimins, W.K. and Shimizu, Y. (1980b), *FEBS Lett.*, **118,** 274–278.

Shimizu, N., Shimizu, Y., Miskimins, W.K., Behzadian, M.A., Kondo, I. and Marwan, M. (1981), *6th International Workshop on Human Gene Mapping,* Abstracts, p. 135.

Shimizu, N., Miskimins, W.K., Gamou, S. and Shimizu, Y. (1982), *Cold Spring Harbor Cell Prolif.*, **9,** 397–402.

Shimizu, N., Shimizu, Y., Fujiki, H. and Sugimura, T. (1983a), *Proc. 14th Princess Takamatsu Cancer Res. Symp.* in press.

Shimizu, N., Shimizu, Y., Fujiki, H. and Sugimura, T. (1983b), *Cancer Res.*, **43,** 4974–4979.

Shimizu, N., Kondo, I., Gamo, S., Behzadian, M.A. and Shimizu, Y. (1983c), *Somat. Cell. Genet.*, in press.

Shimizu, Y. and Shimizu, N. (1980), *Somat. Cell Genet.*, **6,** 583–601.

Shimizu, Y. and Shimizu, N. (1981), *Biochem. Biophys. Res. Commun.* **99,** 536–542.

Shipley, G.D. and Ham, R.G. (1983), *Exp. Cell Res.*, **146,** 249–260.

Shoyab, M., DeLarco, J.E. and Todaro, G.J. (1979), *Nature (London)*, **279,** 387–391.

Sly, W.C. and Fisher, H.D. (1982), *J. Cell. Biochem.*, **18,** 67–85.

Smith, C.J., Wejksnora, P.J., Warner, J.R., Rubin, C.S., and Rosen, O.M. (1979), *Proc. Natl. Acad. Sci. U.S.A.*, **76,** 2725–2729.

Stanley, P. (1980), in *The Biochemistry of Glycoproteins and Proteoglycans* (W.J. Lennarz, ed.), Plenum Press, New York, pp. 161–189.
Stastny, M. and Cohen, S. (1970), *Biochim. Biophys. Acta,* **204,** 578–589.
Stevens, R.L., Schwartz, L.B., Austen, K.F., Lohmander, L.S. and Kimura, J.H. (1982), *J. Biol. Chem.,* **257,** 5745–5750.
Stiles, C.D., Capone, G.T., Scher, C.D., Antoniades, H.N., VanWyk, J.J. and Pledger, W.J. (1979), *Proc. Natl. Acad. Sci. U.S.A.,* **76,** 1279–1283.
Straus, D.S. and Williamson, R.A. (1978), *J. Cell. Physiol.,* **79,** 189–198.
Suzuki, K. and Kono, T. (1980), *Proc. Natl. Acad. Sci. U.S.A.,* **77,** 2542–2545.
Takai, Y., Kishimoto, A., Kikkawa, U., Mori, T. and Nishizuka, Y. (1979), *Biochem. Biophys. Res. Commun.,* **91,** 1218–1224.
Thomas, G., Martin-Perez, J., Seigmann, M. and Otto, A.M. (1982), *Cell,* **30,** 235–242.
Thomopoulos, P., Testa, U., Gourdin, M.-F., Hervy, C., Titeux, M. and Vainchenker, W. (1982), *Eur. J. Biochem.,* **129,** 389–393.
Todaro, G.J., DeLarco, J.E. and Cohen, S. (1976), *Nature (London),* **264,** 26–31.
Todaro, G.J., DeLarco, J.E., Marquardt, H., Bryant, M.L., Sherwin, S.A. and Sliski, A.H. (1979), in *Hormones and Cell Culture* (G. Sato and R. Ross, eds), Cold Spring Harbor Laboratory, New York, pp. 113–127.
Tolleshaug, H., Hobgood, K.K., Brown, M.S. and Goldstein, J.L. (1983), *Cell,* **32,** 941–951.
Tomita, Y.T., Nakamura, T. and Ichihara, A. (1981), *Exp. Cell Res.,* **135,** 363–371.
Trowbridge, I.S. and Omary, M.B. (1981), *Proc. Natl. Acad. Sci. U.S.A.,* **78,** 3039–3043.
Ullrich, A., Shine, J., Chirgwin, J., Pictet, R., Tischer, E., Rutter, W.J. and Goodman, H.M. (1977), *Science,* **196,** 1313–1319.
Ushiro, H. and Cohen, S. (1980), *J. Biol. Chem.,* **255,** 8363–8365.
Van Obberghen, E., Kasuga, M., LeCam, A., Hedo, J.A., Itin, A. and Harrison, L.C. (1981), *Proc. Natl. Acad. Sci. U.S.A.,* **78,** 1052–1056.
Wall, D.A., Wilson, G. and Hubbard, A.L. (1980), *Cell,* **21,** 79–93.
Wang, C.-C., Sonne, O., Hedo, J.A., Cushman, S.W. and Simpson, I.A. (1983), *J. Biol. Chem.,* **258,** 5129–5134.
Wang, J.H. and Waisman, D.M. (1979), *Curr. Top. Cell Regul.* **15,** 47–107.
Waterfield, D., Mayes, E., Stroobant, P., Bennett, P., Young, S., Goodfellow, P., Banting, G. and Ozanne, B. (1982), *J. Cell. Biochem.,* **20,** 149–161.
Waterfield, M.D., Scrace, G.T., Whittle, N., Stroobant, P., Johnson, A., Wasteson, A., Westermark, B., Heldin, C.-H., Huang, J.S. and Deuel, T.F. (1983), *Nature (London),* **304,** 35–39.
Weinstein, I.B., Yamasaki, H., Wigler, M., Lee, L.S., Fisher, P.B., Jeffrey, A.M. and Grunberger, D. (1979), in *Carcinogens Identification and Mechanisms of Action* (A.C. Griffin and R.C. Shaw, eds), Raven Press, New York, pp. 399–418.
Weiss, M.C. and Green, H. (1967), *Proc. Natl. Acad. Sci. U.S.A.,* **58,** 1104–1111.
Wigler, M., Sweet, R., Sim, G.K., Wold, B., Pellicer, A., Lacy, E., Maniatis, T., Silverstein, S. and Axel, R. (1979), *Cell,* **16,** 777–785.
Willingham, M.C. and Pastan, I.H. (1982), *J. Cell Biol.,* **94,** 207–212.
Willingham, M.C., Haigler, H.T., Fitzgerald, D.J.P., Gallo, M.G., Rutherford, A.V. and Pastan, I.H. (1983), *Exp. Cell Res.,* **146,** 163–175.
Wright, J.A., Lewis, W.H. and Parfett, C.L. (1980), *Can. J. Genet. Cytol.,* **22,** 443.
Yang Hu, H.Y. and Aisen, P. (1978), *J. Supramol. Struct.,* **8,** 349–360.

6 Genetics of the Cell Surface of the Preimplantation Embryo: Studies on Antigens Determined by Chromosome 17 in the Mouse

ROBERT P. ERICKSON

EDITOR'S INTRODUCTION

As described in Chapter 1 the complex cellular interactions of the immune system have been successfully analysed by employing the combined power of genetics and antibodies. A similar approach has been attempted for studying other complex interactions of cells, for example in the early mouse embryo. The starting point for these studies was the theoretical argument that the regulatory products of the MHC might have evolved from a more general system which regulated cellular interactions in metazoan development. These theories seemed to be supported by the discovery that the *T/t* locus is linked to *H-2* on chromosome 17 of the mouse. Mutants at the *T/t* locus, isolated from wild populations, display a complex phenotype including segregation distortion in males and homozygous lethality at different times in embryogenesis. In 1967, Robert Erickson and Salome Glucksohn-Waelsch proposed that the pleiotropic effects of the *T/t* locus could be explained by a series of related genes which controlled development through cell surface molecules which in turn regulated cellular interactions.

In Chapter 6 Robert Erickson critically examines the evidence for the expression of chromosome 17 encoded cell surface products in the pre-implantation mouse embryo. He concludes that the hypothesis is not proven and suggests that further technical advances may be needed before manipulative investigation of the cell surface of early embryos is possible.

Acknowledgements

I thank Michael Rosenberg, Amy Moser and Edwin Sanchez for helpful suggestions and Rena Jones for excellent secretarial assistance.

Genetic Analysis of the Cell Surface
(*Receptors and Recognition*, Series B, Volume 16)
Edited by P. Goodfellow
Published in 1984 by Chapman and Hall, 11 New Fetter Lane, London EC4P 4EE

6.1 INTRODUCTION

Preimplantation mammalian embryos have been the object of intensive study during the last two decades. This is not so much because of the inherent fascination of early development as it is because of the ability to remove embryos at these stages from the maternal environment and study them *in vitro*. Thus, the advances that have led to the popularity of this developmental material have been advances in tissue culture techniques and media which have allowed fertilization *in vitro* and culture of the embryo. Most experimental studies do not start with eggs fertilized *in vitro* or the one-cell zygote, which can as easily be flushed from the female reproductive tract as can later stages, but with the two-cell embryo. This is because there is a 'two-cell block' in most strains which makes it difficult to culture an *in-vitro*-fertilized, or a one-cell zygote, through to a blastocyst. However, it is quite easy to culture the two-cell embryo to the blastocyst stage and, with slight changes in media, to have it 'implant' in a culture dish. Thus, the stages that we are primarily concerned about involve the cleavage stages, the 2- to about 16-cell embryo; the morulae stage, a cluster of cells which usually consists of 16 to 32 cells; and the blastocyst stage, which is approximately 64 to 100 cells, and has a blastocoel cavity.

The embryo undergoes several major morphological changes visible at the light microscope level, during this early stage of development, which may be under the direction of cell surface molecules detectable as antigens. The two major and early morphogenetic changes are the processes of compaction and blastulation. Compaction begins at the four- to eight-cell stage, depending on the inbred strain of mice studied when the individual cells (blastomeres) increase cell-to-cell contact, i.e. the membranes of the cells become more juxtaposed to one another. This leaves the embryo with the appearance of a single ball of cells with a smooth outer surface. This process is completed by the morulae stage, when the second most dramatic morphological alteration, blastulation, occurs. Blastulation is the process in which a central cavity or lumen forms, and two morphologically distinct cells appear; an outer layer of cells called the trophoblast, and an inner layer of cells congregated at one end called the inner cell mass. It is the inner cell mass which will give rise to the foetus.

Most of the techniques of modern biochemistry and molecular biology have been applied to the description of these developmental events. Highly sensitive enzyme analyses and two-dimensional gel analyses have revealed much information about changes in synthetic capacities of the embryo and the time of paternal genome activation (as early as the two-cell stage). These tools, and the results obtained with them, have been reviewed many times (see reviews by Van Blerkom and Manes, 1977; Magnuson and Epstein, 1981). There has been much interest in molecular changes in the cell surface of the embryo which might be related to changes in its development. Immunological approaches have been used for this purpose with a wide array of results (reviewed in Solter and Knowles, 1979; Johnson and Calarco, 1980). In order to limit the scope of this chapter, and also to focus on a few examples where the interplay of genetics and development has been probed, I am limiting myself to discussion of antigens, or putative antigens, controlled by genes on chromosome 17. It will become apparent in what follows that the proximal portion of chromosome 17 of the mouse has been heavily implicated in both immunological and developmental events. I shall review the data on expression of major histocompatibility complex (H-2) antigens on the preimplantation embryo, the evidence for antigens determined by the *t*-complex and whether or not they are expressed on the embryo, and evidence that a region of chromosome 17 determines amounts of H-Y, which is apparently expressed on the preimplantation embryo.

6.2 H-2 ANTIGENS ON PREIMPLANTATION EMBRYOS

6.2.1 The antisera used

In exploring the plethora of studies on H-2 antigen expression on preimplantation embryos, it is necessary to first review the evolution of anti-H-2 reagents. *H-2*, the major histocompatibility region of the mouse, was first defined as a red cell alloantigen, i.e. an antigen which could be detected with antisera raised in one inbred strain of mice against tumours or various tissues of other inbred strains (Gorer, 1937). The correlation of red blood cell antigen studies with skin graft rejection studies led to the realization that these red blood cell antigens were present on almost all tissues in the adult mouse (in man, the equivalent antigens are not found on red blood cells) and that their presence evoked strong histoincompatibility reactions. Although skin grafting studies soon disclosed that there were many minor histocompatibility loci, it was, and still is, very difficult to elicit antisera directed to only a single minor histocompatibility locus (Long *et al.*, 1981). Thus, it was usually assumed that the sera obtained by multiple injections of spleen cells from one inbred strain into another inbred strain only contained antibodies to major histocompatibility complex antigens. This type of antiserum was used for early studies of antigenic expression on preimplantation embryos. However, such sera contain antibodies to minor histocompatibility and/or other antigens and are not very specific reagents.

George Snell's major contribution to studies of the major histocompatibility complex was to make congenic lines (see Chapter 1) for this complex (Snell, 1958). This involves preparing an F_1 between two inbred strains differing at the *H-2* locus. The F_1 is backcrossed to the strain (recipient) on which one wishes to place the other *H-2* locus (donor) and selecting offspring heterozygous for the donor H-2 antigens. By repeated crosses, one can move the major histocompatibility complex, and adjoining chromosomal material, from one inbred strain on to the other's background. Although a fairly large chromosomal segment (3–10 centiMorgans; Klein *et al.*, 1982) is moved along with the major histocompatibility complex, one quite rapidly obtains a congenic line with about 99% of its genetic material from the recipient strain and only a small portion from the strain donating the major histocompatibility complex. If one prepares an antiserum in the original recipient strain against the congenic line (of that recipient strain) with a particular *H-2* complex, one obtains a serum directed against the gene products of a much smaller chromosomal region. Such antisera are correctly designated 'anti-H-2'. However, as more detailed genetic studies were made of the major histocompatibility complex, especially by Donald Shreffler and colleagues (Shreffler *et al.*, 1966), who accumulated many mice bearing cross-overs within the major histocompatibility complex, it

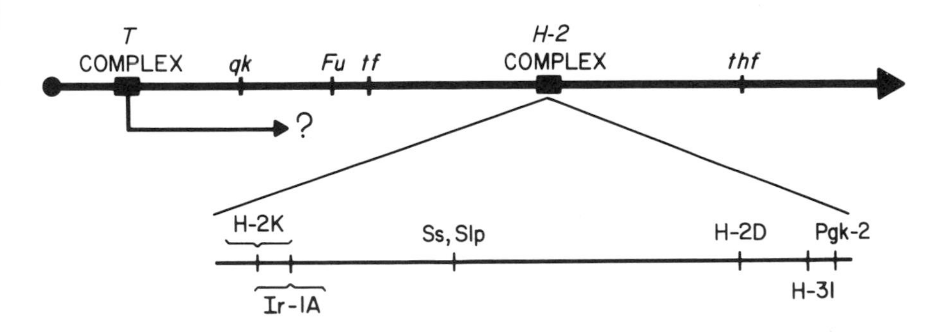

Fig. 6.1 The proximal portion of chromosome 17 of the mouse.

became clear that there are both class I and class II histocompatibility antigens in the complex. The class I antigens are defined as consisting of 45000 molecular weight glycoproteins non-covalently attached to β_2-microglobulin (which maps to chromosome 2; Michaelson, 1981) in the plasma membrane. Class II histocompatibility antigens have chains of 28000 and 32000 molecular weight that are not associated with the β_2-microglobulin. More detailed genetic analysis reveals that class I histocompatibility antigens flank a region of class II histocompatibility antigens (see Fig. 6.1). Using such recombinant mice, it is possible to make antisera to a particular class I antigen, i.e. a D or K or L antigen, or particular class II antigens, i.e. A region, B region, etc. Thus, syngeneic antisera prepared between mice only differing in one subportion of the major histocompatibility complex more accurately define specific antigens. However, such reagents have only recently been applied to studies of the preimplantation embryo.

6.2.2 Classic studies of class I histocompatibility antigen expression on preimplantation embryos

In most studies of H-2 antigen expression on preimplantation embryos, it is assumed that the same array of antigens would be expressed as on adult tissues. Since there are multiple copies of class I loci that are expressed in adult tissues, and many more copies in the genome that can be detected by hybridization of *H-2* cDNA clones (Steinmetz *et al.*, 1981), it is possible that the preimplantation embryo would express class I antigens from alleles not normally expressed in adults. Before this possibility was known, a number of studies reported expression of H-2 antigens on preimplantation embryos (reviewed in Solter and Knowles, 1979). Many of these studies used the first type of antiserum described above which could be contaminated by antibodies to minor histocompatibility loci. The latter are clearly detectable on preimplantation embryos and allowed paternal expression to be detected as early as the eight-cell stage (Muggleton-Harris and Johnson, 1976). When due account is taken of such contaminating specificities, many authors have been unable to find class I H-2 antigens on the preimplantation embryo (Heyner *et al.*, 1969; Palm *et al.*, 1971; Billington *et al.*, 1977). Searle *et al.* (1974) could not detect H-2 antigens on the preimplantation embryo by cell-mediated responses but, when using the morphological technique of immunoperoxidase labelling, have claimed that low levels of H-2 antigens may be present (Searle *et al.*, 1976). Such conflicts and results are frequently due to the confusion of weak non-specific signals (ones that can be obtained from non-immune, control sera on target embryos or from immune sera on non-target embryos, i.e. with embryos which do not contain genes for H-2 antigens against which the antisera are directed) with real signals. Since there is always a subjective element in reading morphological reactions, it is very important that they be read blind.

In the above studies, it is hard to know to what extent such problems have contributed to the different conclusions presented (sometimes from one laboratory: Searle *et al*,. 1974 vs 1976). Cytotoxicity provides a less subjective way of estimating antigen expression on cells since the all-or-none phenomenon of life or death for the cell is more easily scored. Krco and Goldberg (1977) used cytotoxicity to detect paternal H-2 antigens as early as the 8-cell stage. Techniques involving radioisotopes, in which the embryo is counted (by a machine) instead of inspected, provide still more objective techniques. Using an isotope labelled anti-globulin technique, Håkansson *et al.* (1975) could detect paternal H-2 antigens on mouse blastocysts during experimental delay of implantation (the delay, called diapause, occurs when ovariectomy is performed to prevent the oestrogen and progesterone stimulus needed for implantation), but the H-2 antigens disappeared after the onset of implantation. H-2 antigens are not the only cell-surface molecules which appear to be expressed on the diapause, but not the normal, blastocyst: a *Dolichos biflorus* lectin is bound by delayed implanted embryos, and its binding disappears when the embryos are activated for implantation (Chávez and Enders, 1982). Webb *et al.* (1977) performed immunoprecipitation of radiolabelled antigens from embryos grown in media containing isotopically labelled amino acid precursors. They did not find H-2 antigens until the late blastocyst stage when they could precipitate them from the inner cell mass. Using the same techniques, Sawicki *et al.* (unpublished results, cited in Magnuson and Epstein, 1981) were unable to detect any expression or synthesis of H-2 molecules during any stage of preimplantation development. These authors also performed another control which is not always possible but is quite important, i.e. absorbing the antisera with the pure antigen. Heyner and Hunziker (1979) had used an H-2 alloantiserum to detect weak reactivity with the unfertilized oocyte and the one-cell zygote, whereas later stages of preimplantation development were negative. Sawicki *et al.* were able to confirm the presence of weak immunofluorescence on unfertilized eggs with this antiserum but found that the apparent reactivity could not be adsorbed with purified H-2 antigens.

The study of antigens on preimplantation embryos provides several other interesting cases indicating the importance of such controls. Bene and Goldberg (1974) claimed that the sperm-specific isoenzyme of lactate dehydrogenase, LDH-X, could be detected on fertilized, but not on unfertilized, embryos. However, when anti-LDH-X serum was adsorbed by purified LDH-X, the reaction was undiminished (Spielmann *et al.*, 1977). Interestingly, the antibody activity remaining in the rabbit antiserum after adsorption with LDH-X reacted with normally ovulated eggs but not with superovulated eggs. This reactivity could be adsorbed by spleen cells which have no LDH-X, indicating that some other cell surface antigen is involved. Another example is that of chorionic gonadotropin detected on the surface of the preimplantation embryo (Wiley, 1974). This might really be due to a cross-reaction with a serine protease: the reaction of antisera to human lutropin (luteinizing hormone) with oocyte cytoplasm could be adsorbed by a synthetic peptide known to have sequence homology with serine proteases (Jagiello and Mesa-Tejada, 1979). Of course, such problems of cross-reactions of antisera may be obviated by the use of monoclonals but they have their own special problems, especially low avidity (discussed in Erickson *et al.*, 1982).

6.2.3 Are class I histocompatibility antigens that are not normally expressed on adult tissues expressed on preimplantation embryos?

A considerable body of data have recently been accumulated that tumour cells may express unusual class I histocompatibility antigens in place of the normally expected array. Much of the data have been presented at a Workshop which was summarized in an editorial in *Journal of Immunogenetics* (Demant and Festenstein, 1980). There are two major points that need to be discussed in regard to this work. The first is the observation that class I histocompatibility antigens are not expressed in the tightly controlled way that most enzymes are – a mode of genetic control best described by the concept of 'gene dosage'. It is clear that F_1 hybrid mice do not necessarily show half-parental amounts of the parental class I histocompatibility antigens (O'Neill and Blanden, 1979) and viral infection clearly can alter H-2 antigen expression (Meruelo, 1980; Plata *et al.*, 1981). In a similar vein, interferon modulates the amount of HLA class I, but not class II antigen expression (Fellous *et al.*, 1979). Secondly, normally unexpressed copies of class I histocompatibility antigens (of which there are many at the DNA level, varying between different inbred strains of mice; Steinmetz *et al.*, 1981) might be turned on. Evidence for this alternative has been found by Festenstein's group who have shown that a H-2 D^d molecule, as characterized by peptide composition, can be expressed from the *H-2*^{k} genotype of a leukaemia cell line (Festenstein *et al.*, 1979; Schmidt and Festenstein, 1980). Thus, one can ask whether or not a preimplantation mouse embryo (with growth potential which is sometimes compared to that of a tumour), might also express H-2 antigens which are not normally expressed, or modulate antigen expression in a way different from that in adult normal tissues.

Ostrand-Rosenberg *et al.* (1977) found that cultured cell lines derived from mouse blastocysts had variable expression of the H-2 antigen specificities which were expected on the basis of the genotype of the blastocyst. They did not seek H-2 specificities from other genotypes. Cell lines are frequently unstable and these blastocyst-derived cell lines might have been transformed by viruses. However, Cozad and Warner (1981) found that H-2 K^d and D^k molecules were expressed on the blastocyst but H-2 D^b, K^k and D^d molecules were not. In this study, the authors also sought and found expression of H-2 antigens on the embryos which would only be expected in other genotypes. However, as mentioned above, an important control for all such studies is a demonstration that the reactivity can be adsorbed by purified antigen and that has not been done in the latter work. Nonetheless, it remains an intriguing possibility that preimplantation embryos might express atypical H-2 antigens, an hypothesis put forth by Edelman (1975).

6.2.4 Studies of other *H-2* region antigens on preimplantation embryos

The class II (or IA) antigens discussed in Section 6.2.1 have also been sought on the preimplantation embryo. The first studies found them as far back as day 11, but not earlier (Delovitch *et al.*, 1978). Immunofluorescent studies of preimplantation embryos using anti-(class II) reagents were also negative (Heyner and Hunziker, 1981). Interestingly, β_2-microglobulin, which is normally associated with class I histocompatibility-type antigens, is expressed in preimplantation embryos. Sawicki *et al.* (1981) performed immunoprecipitation and two-dimensional gel electrophoresis to detect β_2-microglobulin in preimplantation embryos. Using electrophoretic variants, they were able to demonstrate that the paternal form was expressed as early as the two-cell embryo. However, it is not clear why β_2-microglobulin is expressed. Since β_2-microglobulin is essential for MHC antigen expression on the plasma membrane, it may be in an intracellular pool awaiting the expression of class I antigens during later development and expressed during delayed implantation. Alternatively, the β_2-microglobulin may be necessary for expression of weak transplantation antigens (non-H-2) that are expressed at various times during preimplantation development, for example, H-Y antigen.

Although the overall weight of the evidence is against the expression of either class I or class II antigens on the preimplantation embryo, other than during delayed implantation or, perhaps, on the inner cell mass late in development, it is clear that genetic variation in the H-2 region affects preimplantation development. Blastocysts bearing the H-2^b haplotype, independent of the genetic background on which it has been placed (by making congenic lines), have more cells than do embryos with the H-2^k haplotype (Goldbard *et al.*, 1982). This is due to a more rapid rate of cell division. These differences in growth rate could well be due to some *H-2*-linked gene not expressed as an antigen on the surface of embryos. Candidate variations would include the transmembrane methyltransferase I, an enzyme affecting cell membrane fluidity by methylating phosphatidylethanolamine to phosphatidylcholine, a component of which is linked to *H-2* (Markovac and Erickson, unpublished work). Whatever the basis of the association, it remains an intriguing one which should encourage further studies, hopefully at the mRNA level using DNA clones, of *H-2*-related expression in preimplantation embryos.

6.3 PUTATIVE *T*-COMPLEX-DETERMINED ANTIGENS AND THE PREIMPLANTATION EMBRYO

6.3.1 An attempt to simplify the *t*-complex

A portion of chromosome 17, near the centromere, can exist in an unusual condition. The most essential feature of this condition is its inability to crossover when a wild-type chromosome is across from it. Such an unusual 'patch' of chromosome is termed a *t*-allele. They were originally found in wild populations of mice but can be maintained in the laboratory as balanced lethal lines. Because of the cross-over suppression, a '*t*-allele' acts as a point mutation. It is only in recent years that it has become clear that the 'patch' can extend for as much as 15 centiMorgans. This was deduced using deletions that extend into this region (Lyon and Bechtol, 1977; Erickson *et al.*, 1978) and in situations where cross-overs can occur between two *t*-alleles which are 'across' from each other (Silver and Artzt, 1981). Thus, a large number of genetic attributes should be encoded by this amount of DNA. A second, unusual feature of these *t*-alleles is the ability to distort transmission ratios. A male, but not a female, carrying a *t*-allele may transmit it to 95%, or even 98%, of its progeny. This property is also unusual in that genetic dissection of this region discloses that transmission ratio distortion depends on several areas of the chromosome (Lyon and Mason, 1977; Hammerberg, 1982a). The transmission ratio distortion allows *t*-alleles to be maintained in the population despite homozygous lethality which is also present in most *t*-alleles. It is the last property which has attracted the most attention to *t*-alleles. The lethal genes exist as a series of mutations which affect developmental processes at different stages, from ones arresting development as early as the preimplantation embryo (t^{12}) to as late as mid-gestation. Finally, and probably least important conceptually, is the ability of these two alleles to interact with a common laboratory mutation, *T*, which by itself causes a short-tail, to result in a tail-less mouse. This property is very useful for 'keeping track' of *t*-alleles in genetic crosses.

Much of the work on the embryology of *t*-alleles, including some of this author's, has proceeded on the assumption that the embryological lethal effects would be interrelated, i.e. when one thought of this as a very small chromosomal segment, the different developmental arrests were thought to be different manifestations of some single developmental pathway. Now that it is clear that these developmental mutations can be spread out over as much as 15 centiMorgans of chromosome, there seems less reason to interrelate them pathophysiologically. In the era of attempting to find a unifying explanation for the embryonic lethality, the hypothesis that these mutations affected cell surface molecules involved in development seemed a reasonable one. There have been multiple attempts to discover such antigens. In what follows, we will briefly discuss the status of the search.

6.3.2 F9 embryonal carcinoma antigen and t^{12}

One antigen that is expressed in early embryos and that has been reported to have a role in morphogenesis is the F9 teratocarcinoma antigen. The class of teratocarcinoma antigens which cross-react with embryos was initially found by Edidin and co-workers who raised antisera in rabbits directed against mouse teratocarcinoma (Edidin *et al.*, 1971). Artzt and co-workers produced a syngeneic antiserum against cell lines from such a teratocarcinoma using irradiated cells from a nullipotent embryonal carcinoma cell line, F9 (Artzt *et al.*, 1973). This antiserum (anti-F9) also reacted with preimplantation embryos and male germ line cells. Developmental lethal mutations of the *t*-complex of the mouse were reported to alter F9 antigen activity (Artzt *et al.*, 1974). It was suggested that the structural gene for the F9 antigen was located in the T region of chromosome 17 of the mouse since (1) twice as many sperm cells from a $+/t^{w32}$(t^{w32} is like t^{12}) mouse were needed to remove a standard amount of F9 antiserum when compared to sperm from a +/+ mouse (Artzt *et al.*, 1974) and (2) expression of the F9 antigen appeared to segregate among the progeny of crosses producing t^{w32}/t^{w32} or t^{w5}/t^{w5} embryos (Kemler *et al.*, 1976). The number of negative embryos approximated the expected number of *t*-allele homozygotes. Originally this group reported that t^{12} homozygous embryos expressed no F9 antigen but this claim has since been withdrawn (Marticorena *et al.*, 1978). A thorough study of the expression of the F9 antigen, using several different antisera, disclosed no deficiency in reactivity with a variety of embryos, including t^{12}/t^{12} (Erickson and Lewis, 1980). To the extent that a functional role for F9 antigen has been found in preimplantation embryos, it seems that it may have a role in compaction, the process by which the separated blastomeres become fused to make a smooth surface (see the introduction), since anti-F9 monovalent fragments block this step (Kemler *et al.*, 1977). However, this is a developmental stage which occurs in a normal fashion in t^{12}/t^{12} embryos, so it would be hard to relate deficiencies in F9 antigen to the stage of developmental arrest of t^{12}/t^{12} embryos.

6.3.3 Putative t^n antigens

The search for cell surface antigens determined by *t*-alleles has mostly focused on spermatozoa, since a uniform hypothesis of action would lead one to expect that such antigens, if they existed, would be expressed on spermatozoa (because of the male transmission ratio distortion) as well as on preimplantation embryos. Evidence for the presence of an altered sperm-associated antigen determined by the dominant *T* allele was found with a mouse antiserum raised against spermatozoa from mice heterozygous for *T* (Bennett *et al.*, 1972). However, the relevance of this spermatozoal antigen to the physiology of the *t*-complex is questionable as *T* is silent in terms of sperm function. Further studies claimed partially distinctive antigens associated with several different recessive *t*-alleles, and produced evidence compatible with post-meiotic expression of a putative *t*-antigen in differentially viable *t* and wild-type spermatozoa (Yanagisawa *et al.*, 1974a,b). Studies of antibodies to spermatozoa from animals bearing various *t*-alleles, as well as analysis of the cross-reacting components, suggested that a complex antigenic system might be involved (Artzt and Bennett, 1977). Recent extensive experiments have been unable to confirm the existence of such putative *t*-antigens on spermatozoa (Goodfellow *et al.*, 1979; Gable *et al.*, 1979).

On the other hand, any other large chromosomal region could include genes for several cell-surface proteins. Loci coding for minor histocompatibility antigens are found in, or near, the *t*-complex (Flaherty, 1975; Artzt *et al.*, 1977). Hammerberg (1982b) has described an antigen only detected by cell-mediated lymphocytotoxicity which all the *t*-alleles studied shared. This might be an allelic form of one of the minor histocompatibility loci described above or might represent an immunological detection of p63/6.9, a cell surface protein variant found with all t^{n} alleles (Silver *et al.*, 1979).

6.4 A LOCUS DETERMINING THE AMOUNT OF H-Y ANTIGEN AND CHROMOSOME 17

H-Y, or histocompatibility Y, and its putative role in male sex determination is discussed in Chapter 7. Krco and Goldberg (1976) first reported that H-Y antigen was expressed on preimplantation embryos – they found that about one-half of preimplantation embryos could be killed by anti-(H-Y) sera. However, it was only an inference that these were Y chromosome containing embryos, since no studies were made of the surviving embryos (Erickson, 1977). Epstein *et al.* (1980) cultured the embryos surviving anti-(H-Y) treatment (about 50%) and found 98% of them to be XX. Thus, it seems that the Y chromosome leads to differential antigen expression by the eight-cell stage. One gene system, *Hye*, is thought to regulate the level of H-Y. This locus was defined as governing H-Y 'antigenic strength' in the popliteal node enlargement assay, a test system which measures a cell-mediated response (Králová and Demant, 1976). The effects of this locus have been mapped somewhat nearer to the *T* locus than the *H-2* complex on chromosome 17 (Králová and Lengerová, 1979). This difference in H-Y levels determined by chromosome 17 has been confirmed serologically (Shapiro and Erickson, 1984). In the latter work, it was found that the t^{12}-haplotype carried an *Hye* low allele. Thus, there is a possible relationship between the *Hye* locus and the apparent deficiency of F9 antigen reported for the t^{12} allele. F9, much like H-Y (Shapiro and Erickson, 1981) is an antigen whose serological recognition is thought to involve a polysaccharide structure that contains galactosyl moieties (Muramatsu *et al.*, 1979; Prujansky-Jacobovits *et al.*, 1979). Thus, chromosome 17 determinants of cell surface glycosyltransferases (Shur and Bennett, 1979) may affect several antigens of preimplantation embryos. Our laboratory, however, using a different acceptor for galactosyltransferase, was not able to find any strain-specific differences in adult liver galactosyltransferase (Shapiro, 1982). In the original report (Shur and Bennett, 1979), larger differences were found among some of the normal inbred strains that were used as controls than were found related to *t*-alleles. Thus, this too, must be considered only a possible chromosomal-17-determined change in cell surface properties.

6.5 SUMMARY

Despite intensive investigation during the last several decades, one must conclude that antigenic changes in cell surfaces of preimplantation embryos determined by chromosome 17 remain interesting possibilities which are not yet instructive for our understanding of development. Immunological reagents are attractive tools because of their tremendous sensitivity. However, as illustrated in a number of examples presented in this chapter, they frequently do not have adequate specificity and many misleading results can be obtained. Monoclonal antibodies are the current immunological tools of great interest but, as already mentioned, they too have their problems. When we consider the problem of studying the preimplantation embryo, one conceptual problem may be the biological scientist's view of it as a 'simple' organism. Perhaps it is better to view this stage in terms of the number of developmental days that it spans rather than in terms of the number of cells it possesses. A general contrast is often made between the finding of paternal genome expression as early as the two-cell stage in the mammalian embryo while amphibian or sea-urchin embryos manage quite well with only the maternal contribution until a much later stage of development. The contrast is based upon the perceived differences in morphological complexity rather than the time scale. It takes 24 hours for a mammalian embryo to get to the two-cell stage while, in 24 hours, a sea-urchin embryo already has mesenchyme cells. At the time that a mammalian embryo has just reached the blastocyst stage, a sea urchin embryo is a feeding pluteus. Thus, further refinements in microtechniques, perhaps especially involving DNA clones for studies of specific messages, may help us eventually to approach a degree of analysis of the mammalian preimplantation embryo that has already been achieved with the sea-urchin embryo, for instance, where much larger quantities of material are available.

REFERENCES

Artzt, K. and Bennett, D. (1977), *Immunogenetics*, **5**, 97–107.

Artzt, K., Dubois, P., Bennett, D., Condamine, H., Babinet, C. and Jacob, F. (1973), *Proc. Natl. Acad. Sci. U.S.A.*, **70**, 2988–2992.

Artzt, K., Bennett, D. and Jacob, F. (1974), *Proc. Natl. Acad. Sci. U.S.A.*, **71**, 811–814.

Artzt, K., Hamburger, L. and Flaherty, L. (1977), *Immunogenetics*, **5**, 477–480.

Bene, M. and Goldberg, E. (1974), *J. Exp. Zool.*, **189**, 261–266.

Bennett, D., Golberg, E., Dunn, L.C. and Boyse, E.A. (1972), *Proc. Natl. Acad. Sci. U.S.A.*, **69**, 2076–2080.

Billington, W.D., Jenkinson, E.J., Searle, R.F. and Sellens, M.H. (1977), *Transplant. Proc.*, **9**, 1371–1377.

Chávez, D.J. and Enders, A.C. (1982), *Biol. Reprod.*, **26**, 545–552.

Cozad, K.M. and Warner, C.M. (1981), *J. Exp. Zool.*, **218**, 313–320.

Delovitch, T.L., Press, J.L. and McDevitt, H.O. (1978), *J. Immunol.*, **120**, 818–824.

Demant, P. and Festenstein, H. (1980), *J. Immunogenet.*, **7**, 1–6.

Edelman, G.M. (1975), in *The Cell Surface, Immunological and Chemical Approaches* (B. Kahan and R. Reisfeld, eds), Plenum Press, New York, pp. 260–266.

Edidin, M., Patthey, H.L., McGuire, E.J. and Sheffield, W.D. (1971), in *Embryonic and Fetal Antigens in Cancer* (N.G. Anderson and T.H. Coggin, eds), Technical Information Bulletin, Atomic Energy Commission, Washington D.C., pp. 239–247.
Epstein, C.J., Smith, S. and Travis, B. (1980), *Tissue Antigens,* **15,** 63–67.
Erickson, R.P. (1977), *Nature (London),* **265,** 59–61.
Erickson, R.P. and Lewis, S.E. (1980), *J. Reprod. Immunol.,* **2,** 293–304.
Erickson, R.P., Lewis, S.E. and Slusser, K.S. (1978) *Nature (London),* **274,** 163–164.
Erickson, R.P., Kay, G., Hewett-Emmett, D., Tashian, R.E. and Claflin, J.L. (1982), *Biochem. Genet.,* **20,** 809–819.
Fellous, M., Kamoun, M., Gresser, I. and Bono, R. (1979), *Eur. J. Immunol.,* **9,** 446–492.
Festenstein, H., Schmidt, W., Testorelli, C., DiGiorgi, L., Morelli, O., Matossian-Rogers, A. and Atfield, G. (1979), *J. Immunogenet.,* **6,** 263–270.
Flaherty, L. (1975), *Immunogenetics,* **2,** 325–329.
Gable, R.J., Levinson, J.R., McDevitt, H.O. And Goodfellow, P.N. (1979), *Tissue Antigens,* **13,** 177–185.
Goldbard, S., Verbanac, K.M. and Warner, C.M. (1982), *J. Immunogenet.,* **9,** 77–82.
Goodfellow, P.N., Levinson, J.R., Gable, R.J. and McDevitt, H.O. (1979), *J. Reprod. Immunol.,* **1,** 11–21.
Gorer, P.A. (1937), *J. Pathol. Bacteriol.,* **44,** 691–697.
Håkansson, S., Keyner, S., Sundgkvist, K.-G. and Beryström, S. (1975), *Int. J. Fertil.,* **20,** 137–140.
Hammerberg, C. (1982a), *Genet. Res.,* **39,** 219–226.
Hammerberg, C. (1982b), *J. Immunogenet.,* **9,** 179–184.
Heyner, S. and Hunziker, R.D. (1979), *Dev. Genet.,* **1,** 69–76.
Heyner, S. and Hunziker, R.D. (1981), *J. Immunogenet.,* **8,** 523–528.
Heyner, S., Brinster, R.L. and Palm, J. (1969), *Nature (London),* **222,** 783–784.
Jagiello, G. and Mesa-Tejada, R. (1979), *Endocrinology,* **104,** 302–307.
Johnson, L.V. and Calarco, P.G. (1980), *Anat. Rec.,* **106,** 201–219.
Kemler, R., Babinet, C., Condamine, H., Gachelin, G., Guenet, J.L. and Jacob, F. (1976), *Proc. Natl. Acad. Sci. U.S.A.,* **73,** 4080–4084.
Kemler, R., Babinet, C., Eisen, H. and Jacob, F. (1977), *Proc. Natl. Acad. Sci. U.S.A.,* **744,** 4449–4452.
Klein, D., Tewarson, S., Figueroa, F. and Klein, J. (1982), *Immunogenetics,* **16,** 319–326.
Králová, J. and Demant, P. (1976), *Immunogenetics,* **3,** 583–594.
Králová, J. and Lengerová, A. (1979), *J. Immunogenet.,* **6,** 429–438.
Krco, C.J. and Goldberg, E.H. (1976), *Science,* **193,** 1134–1135.
Krco, C.J. and Goldberg, E.H. (1977), *Transplant. Proc.,* **9,** 1367–1370.
Long, P.M., Lafuse, W.P. and Davis, C.S. (1981), *J. Immunol.,* **127,** 825–830.
Lyon, M.F. and Bechtol, K. (1977), *Genet. Res.,* **30,** 63–76.
Lyon, M.F. and Mason, I. (1977), *Genet. Res.,* **29,** 255–266.
Magnuson, T. and Epstein, C.J. (1981), *Biol. Rev.,* **56,** 369–408.
Marticorena, P., Artzt, K. and Bennett, D. (1978), *Immunogenetics,* **7,** 337–347.
Meruelo, D. (1980), *J. Immunogenet.,* **7,** 81–90.
Michaelson, J. (1981), *Immunogenetics,* **13,** 167–171.

Muggleton-Harris, A.L. and Johnson, M.H. (1976), *J. Embryol. Exp. Morphol.*, **35**, 59–72.

Muramatsu, T., Gachelin, G. and Jacob, F. (1979), *Biochim. Biophys. Acta*, **587**, 392–406.

O'Neill, H.C. and Blanden, R.V. (1979), *J. Exp. Med.*, **149**, 724–731.

Ostrand-Rosenberg, S., Hammerberg, C., Edidin, M. and Sherman, M.I. (1977), *Immunogenetics*, **4**, 127–136.

Palm, J., Heyner, S. and Brinster, R.L. (1971), *J. Exp. Med.*, **133**, 1282–1293.

Plata, F., Tilkin, A.-F., Lévy, J.-P. and Lilly, F. (1981), *J. Exp. Med.*, **154**, 1795–1810.

Prujansky-Jacobovits, A., Gachelin, G., Muramatsu, T., Sharon, N. and Jacob, F. (1979), *Biochem. Biophys. Res. Commun.*, **89**, 448–455.

Sawicki, J.A., Magnuson, T. and Epstein, C.J. (1981), *Nature (London)*, **294**, 450–451.

Schmidt, W. and Festenstein, H. (1980), *J. Immunogenet.*, **7**, 7–17.

Searle, R.F., Johnson, M.H., Billington, W.D., Elson, J. and Clutterbuch-Jackson, S. (1974), *Transplantation*, **18**, 136–150.

Searle, R.F., Sellens, M.H., Elson, J., Jenkinson, E.J. and Billington, W.D. (1976), *J. Exp. Med.*, **143**, 348–359.

Shapiro, M. (1982), Ph.D. Thesis, University of Michigan, Ann Arbor.

Shapiro, M. and Erickson, R.P. (1981), *Nature (London)*, **290**, 503–505.

Shapiro, M. and Erickson, R.P. (1984), *J. Reprod. Immunol.* (in press).

Shreffler, D.C., Amos, D.B. and Mark, R. (1966), *Transplantation*, **4**, 300–322.

Shur, B.D. and Bennett, D. (1979), *Dev. Biol.*, **71**, 243–259.

Silver, L. and Artzt, K. (1981), *Nature (London)*, **290**, 68–70.

Silver, L.M., Artzt, K. and Bennett, D. (1979), *Cell*, **17**, 275–284.

Snell, G.D. (1958), *J. Natl. Cancer Inst.*, **21**, 843–877.

Solter, D. and Knowles, B.B. (1979), *Curr. Top. Dev. Biol.*, **13**, 139–166.

Spielman, H., Eibs, H.-G., Mentzel, C. and Nagel, D. (1977), *J. Reprod. Fertil.*, **50**, 47–52.

Steinmetz, M., Frelinger, J.G., Fisher, D., Hunkapiller, T., Pereira, D., Weissman, S.M., Uehara, H., Nathanson, S. and Hood, L. (1981), *Cell*, **24**, 125–134.

Van Blerkom, J. and Manes, C. (1977), in *Concepts in Mammalian Embryology* (M.I. Sherman, ed.), M.I.T. Press, Cambridge, pp. 37–94.

Webb, C.G., Gall, W.E. and Edelman, G.M. (1977), *J. Exp. Med.*, **146**, 923–932.

Wiley, L.D. (1974), *Nature (London)*, **252**, 715–716.

Yanagisawa, K., Bennett, D. Boyse, E.A., Dunn, L.C. and Dimeo, A. (1974a), *Immunogenetics*, **1**, 57–67.

Yanagisawa, K., Pollard, D.R., Bennett, D., Dunn, L.C. and Boyse, E.A. (1974b), *Immunogenetics*, **1**, 91–96.

7 The Male-Specific Antigen (H-Y) and Sexual Differentiation

PETER W. ANDREWS

EDITOR'S INTRODUCTION

Female mammals have two X chromosomes, males have one X and one Y chromosome. Individuals with abnormal sex chromosome constitutions are phenotypically male if they have a Y chromosome and female if they lack a Y chromosome. These observations suggest that maleness is due to a gene or genes on the Y chromosome. Because the mammalian Y chromosome is generally bereft of genetic markers, any gene found to reside on the Y chromosome is worth considering as a candidate for sex determination.

Male and female mice from the same inbred strain differ genetically only by the Y chromosome, yet such females reject male skin grafts. This was the first demonstration of the Y-controlled cell surface antigen, or antigens, H-Y. Cross-immunization of females with male cells can also be used to produce male specific 'T cell killers' and antibodies. The relationship between H-Y antigens defined by transplantation, T cell killing and serological assays is the subject of the review by Peter Andrews which constitutes Chapter 7. Also considered is the role of these various H-Y antigens in sex determination. It is suggested that a crucial test for the hypothesis that H-Y antigen is the male sex-determining substance has not yet been devised.

Acknowledgements

This work was supported, in part, by a grant, CA-29894, from the United States Public Health Service.

Genetic Analysis of the Cell Surface
(*Receptors and Recognition*, Series B, Volume 16)
Edited by P. Goodfellow
Published in 1984 by Chapman and Hall, 11 New Fetter Lane, London EC4P 4EE

7.1 INTRODUCTION

In 1932 Bittner reported a difference between male and female mice in their susceptibility to a transplantable tumour, and presciently suggested that a Y-chromosomal factor might be involved. Also in the same year, Kozelka (1932) noted in chickens a higher incidence of graft rejection when female skin was grafted on to male hosts than when grafts were made between other sex combinations. However, the experimental study of what, in mammals, became known as the male-specific transplantation antigen, or histocompatibility antigen-Y (H-Y, Billingham and Silvers, 1960) begins with the observations of Eichwald and Silmser (1955). They reported that in experiments with the C57BL and A/Jax inbred strains of mice none of 11 C57BL male to C57BL female grafts, and only 10 of 18 A/Jax male to A/Jax female grafts survived, whereas intrastrain male to male, female to female and female to male grafts almost always survived. Hauschka (1955) and Snell (1956), in discussing these observations, both suggested that they would be readily explained by a Y-chromosome-linked transplantation antigen. As we shall discuss, a sex-limited transplantation antigen (or antigens) of this type (sometimes expressed in males, sometimes in females) is widely distributed, not only in mammals, but in other classes of vertebrates as well.

Among the many minor, or weak, transplantation antigens of the mouse, H-Y has been especially popular with immunologists as a model for investigating the workings of the immune system. The ease with which H-Y can be studied (males and females of any inbred strain provide a ready-made co-isogenic pair), the genetic variations in its immunogenicity and in responsiveness to it, and the relative ease with which tolerance can be induced have all contributed to its usefulness. There are also possible clinical implications for organ transplantation in humans (Goulmy *et al.*, 1978; Oliver, 1974). However, in recent years wider interest in the H-Y antigen has been generated by the suggestion that in mammals, during embryogenesis, it exercises a crucial causative role in the differentiation of the primitive gonad into a testis rather than an ovary (Wachtel *et al.*, 1975a). In species, such as birds, in which females are the heterogametic sex and express this antigen, it was suggested to cause ovarian rather than testicular differentiation. This proposition is particularly attractive in the light of the idea that cell–cell interactions mediated by cell-surface molecules generally play an important part in regulating cellular differentiation and morphogenesis during embryonic development (Bennett *et al.*, 1972). The idea that the H-Y antigen acts as the primary sex-determining molecule which promotes testicular or ovarian differentiation depending upon the species, however, has also generated considerable controversy. Although widely debated, the argument has not yet been resolved (e.g. Goodfellow and Andrews, 1982; Simpson, 1982a). In part this is because of the multitude of assays which has been employed, resulting in the possibility (or even probability?) that different authors have been studying different molecules. Further, the various available serological assays, potentially the most useful for biochemical and functional studies, are also technically difficult to conduct, difficult to control, and, for lack of sufficient quantities of standardized reagents, not readily amenable to proper quantification.

The object of this chapter is to attempt to collate and assess the various data and points of view which bear on the postulated function of the H-Y antigen in sexual differentiation. To do this it is necessary to review in some detail the salient features of the immunology of this H-Y antigen, but for a more extensive treatment of this aspect, the reader is referred elsewhere (Gasser and Silvers, 1972; Silvers and Wachtel, 1977; Simpson, 1982b; Billingham and Hings, 1981). I have already alluded to the possibility that different authors using different assays may be studying diverse molecules. To avoid prejudicing whether or not the male-specific antigens detected by the three main groups of techniques that have been used (namely, transplantation, cell-mediated cytotoxicity and serology) are the same, I shall modify the symbol H-Y, where distinction may be necessary, with a subscript and refer to H-Y_T, H-Y_C and H-Y_S, respectively. I shall continue to use the term H-Y, unmodified, whenever it is useful to discuss this sex-limited antigen (or antigens) in general terms, however assayed. Some authors have proposed that 'H-Y' be reserved for the antigen detected by transplantation, and other T-cell-dependent assays, whereas the 'serologically defined male-specific antigen' should be designated 'SDM' (Silvers *et al.*, 1982). While there may be some merit in this, the already widespread use of the term H-Y to refer to both classes of antigen might mean that this revision of the terminology could be confusing. Further, it is not certain that the various T-cell-dependent assays are detecting only one antigen. The use of subscripts permits greater flexibility in nomenclature, and they can be dropped if and when antigens identified by two different assays are shown to be identical. Much of the experimental work concerning H-Y antigen has been conducted in laboratory mice. Throughout this chapter it should be assumed that the results being discussed refer to laboratory mice unless a different species is indicated.

7.2 TRANSPLANTATION H-Y (H-Y_T)

7.2.1 Expression and response

The observation of Eichwald and Silmser (1955) contradicted the laws of transplantation, as they were understood at that time, namely that between genetically identical individuals, as between identical twins or members of a highly inbred strain, there should be complete acceptance of transplanted tissues (Little, 1941). However, the rejection of male skin by syngeneic female mice had all the hallmarks of an immunological reaction: females that had previously rejected male skin rejected subsequent grafts more rapidly (second-set response) (Eichwald *et al.*, 1957, 1958; Krohn 1958; Sachs and Heller, 1958); tolerance could be induced by neonatal exposure to male skin (Billingham and Silvers, 1960); and finally the histology of grafts undergoing rejection was similar to that seen in allogeneic graft rejections (Krohn, 1958; Eichwald and Lustgraaf, 1961). On the other hand, survival of the male grafts did not merely depend on the host's hormonal environment, since castrated and testosterone-treated females continued to reject male grafts, whereas castrated and oestrogen-treated male hosts did not (Eichwald *et al.*, 1957, 1958; Bernstein *et al.*, 1958). Thus Billingham and Silvers (1960) suggested that a male-specific minor histocompatibility antigen was the basis for the male-female incompatibility and they proposed the term 'histocompatibility antigen-Y', or H-Y, in recognition of the simplest hypothesis that this antigen was Y-linked (Hauschka, 1955; Snell, 1956). Although most easily detected by skin grafts, it soon became evident, both by orthotopic and ectopic transplants of other organs and by their ability to presensitize females to subsequent male skin grafts, that the H-Y_T transplantation antigen is expressed by a wide variety of cell types. These include lymphoid tissues, such as thymus, lymph nodes, spleen and blood cells (Hirsch, 1957; Feldman, 1958; Eichwald *et al.*, 1958; Sachs and Heller, 1958), endocrine tissues, such as thyroid, parathyroid, adrenal cortex, islet cells and pituitary (Gittes and Russell, 1961; Naji *et al.*, 1981; Hoshino and Moore, 1968), heart (Judd and Trentin, 1971), lung and salivary gland (Eichwald *et al.*, 1958) and liver (Eichwald *et al.*, 1958; Sachs and Heller, 1958; Feldman, 1958).

In some strains of mice, notably C57BL/6 and C57BL/10, females almost always reject syngeneic male skin grafts, often within 20–40 days. However, in other strains the response may be much more variable, with longer median rejection times and with only some females responding; sometimes rejection is only observed after presensitization. In yet other strains no rejection of male grafts by females is ever observed, even after presensitization. Nevertheless, males of non-rejector strains do express H-Y_T antigen since F_1 hybrid females between a rejector and non-rejector strain will reject skin from males and not females of either parental strain or the F_1 hybrids (Bernstein *et al.*, 1958; Eichwald *et al.*, 1957, 1958; Zaalberg, 1959). Further, the injection of cells from males but not females of non-rejector strains can presensitize females of rejector strains to subsequent male skin grafts (Celada and Welshons, 1963), or render them tolerant (Billingham and Silvers, 1960) depending upon the protocol used. Since the immunological phenomena of presensitization and tolerance are highly specific, these data indicate not only that males of all the strains of mice tested express H-Y_T antigen, but also that it is immunologically identical in all strains. These same data also eliminate the possibility that, in general, non-rejector females fail to respond to male skin grafts because they also express H-Y_T antigen and so are tolerant to it (Michie and McLaren,

1958). However, Hauschka and Holdridge (1962) did report data suggesting that non-rejector females of one strain of mice ('Yellow') did express H-Y_T antigen, but this exception has not been pursued.

The ability of females to reject male grafts is evidently a dominant character since female F_1 hybrids between rejector and non-rejector strains also reject male grafts. From backcrosses to the non-rejector parental strain, it was inferred that at least two autosomal loci are involved (Klein and Linder, 1961; Eichwald and Wetzel, 1965). Further study has shown that at least one of these 'response' genes is *H-2*-linked and located in the major histocompatibility complex (MHC) (Bailey and Hoste, 1971; Gasser and Silvers, 1971a,b; Stimpfling and Reichert, 1971), and more specifically in the *IB* subregion (Hurme *et al.*, 1978a). Thus females carrying the IB^b allele are generally able to reject syngeneic male skin grafts. Some female lacking the *b* allele at this locus are also able to reject male grafts, but this ability is thought to be a function of other loci which are not linked to *H-2* (Bailey, 1971; Gasser and Silvers, 1971b; Gasser and Shreffler, 1974).

The influence of the MHC on the H-Y system is not limited to the female response, but also extends to the immunogenicity of the male antigen. For example, using the *H-2* congenic pair B10 ($H\text{-}2^b$) and B10.BR ($H\text{-}2^k$), Wachtel *et al.* (1973a,b) showed that B10.BR male skin was rejected significantly faster than B10 male skin by (B10×B10.BR)F_1 hybrid females. It was suggested that H-Y_T is more immunogenic when present with $H\text{-}2^k$ than when present with $H\text{-}2^b$. Králová and Démant (1976) confirmed this and showed that it was the *K* end of the *H-2* region which determines the immunogenicity of H-Y. Interestingly, the effects of these two *H-2* haplotypes on H-Y immunogenicity are the opposite of their effects upon female responsiveness to H-Y; $H\text{-}2^b$ females generally respond much more strongly to H-Y than do $H\text{-}2^k$ females (Gasser and Silvers, 1971a,b).

7.2.2 Y-linkage or a sex-limited trait

The question of whether the H-Y antigen is specified by a Y-chromosome-linked gene, or whether it is specified by the X chromosome or an autosome but expressed only in males, is a long-standing problem which also bears on the issue of whether H-Y antigen is involved in regulating sexual differentiation. Many male 'secondary sexual characteristics' are regulated by the male hormones testosterone and dihydrotestosterone, produced primarily in the testis (see Ohno, 1971). Accordingly, many experiments have attempted to eliminate H-Y in males by castration, and to induce it in females by ovariectomy or by exposure to male hormones. In early experiments of this type, Bernstein *et al.* (1958) and Eichwald *et al.* (1957, 1958) noted that ovariectomized females rejected male grafts as fast as control females. Testosterone treatment of the females and castration of the male donors also had no effect. However, Vojtíšková and Poláčková (1966) and Poláčková and Vojtíšková (1968) found that grafts from male mice that had been castrated within 12 h of birth frequently did survive on female hosts. In this case the mean survival time of the grafts and the proportion surviving permanently were similar to the results obtained by grafting neonatal male skin, which is less immunogenic than adult male skin, and which may indeed induce tolerance to H-Y_T (Billingham *et al.*, 1965). It seemed, therefore, that neonatal castration arrests development of H-Y_T immunogenicity in its neonatal state, apparently in a specific way since no such effect was noted with another minor histocompatibility antigen, H-3. Nevertheless, H-Y_T antigen is evidently expressed in these neonatally castrated males since they do not reject normal adult male skin. Also H-Y_T can be detected in foetal mice as early as 11 days of gestation (Poláčková, 1970). Thus male castration appears to affect the immunogenicity of the H-Y_T antigen, rather than to prevent its expression. However, Weissman (1973) did not observe this effect, finding that skin from neonatally orchidectomized males was rejected as fast as skin from normal males.

To examine further whether a male hormonal environment can induce H-Y antigen in female cells, Engelstein (1967) passaged female skin on male hosts before transferring it back to syngeneic females, which often rejected it. Control experiments suggested that rejection was not due to contamination of the grafts with male tissue from the intermediate host, and Engelstein concluded that H-Y_T antigen had been induced in the female grafts. However, contrary results were obtained in similar experiments by Silvers *et al.* (1968) and Poláčková (1969), who found that contaminating tissue from the intermediate host can be important in influencing the results. They concluded that H-Y_T antigen is not inducible in female cells by a male hormone environment. On balance, and in view of the complex nature of the experiment, the latter conclusion is most likely correct.

A genetic experiment to test the dependence of H-Y_T expression on the male hormonal environment was provided by the X-linked mutant, *tfm* (*testicular feminization*). Individuals who have an XY chromosomal constitution but also carry the *tfm* mutation develop as phenotypic females with testes instead of ovaries (Lyon and Hawkes, 1970). Ohno (1971), arguing that all male secondary sexual characteristics normally develop as a result of interaction of testosterone, produced in the testes, with a receptor in the target cells, has shown the *tfm*, X/Y female mice lack a cytoplasmic binding protein for testosterone. Nevertheless, these same 'females' do express H-Y_T antigen (Bennett *et al.*, 1975). This suggests that H-Y_T expression is independent of testosterone, although Erickson (1977) has pointed out that testosterone may sometimes act via mechanisms other than those involving the *tfm*-controlled testosterone-binding protein.

Since most of the available data support the view that the presence or absence of H-Y_T is independent of the sex hormones, although they may modulate its immunogenicity in a quantitative way, many attempts have been made to demonstrate that it is specified by a Y-linked gene. This would be simple if polymorphic variants could be found but, despite some claims to the contrary, none have been convincingly demonstrated. To test for variants several authors studied the grafting of male F_1 hybrid mice (e.g. [C57BL/6♀×A♂]F_1♂) with male skin from the maternal parent strain (in this case C57BL/6♂) so that the only genetic component of the donor skin lacking in the recipient is the Y chromosome. If, in this example, C57BL and A strain males express immunologically different Y-chromosome-encoded antigens, graft rejection might result. Hildemann and Cooper (1967) and Hildemann *et al.* (1970) did report such a result and suggested that A strain and C57BL/6 mice differed in a Y-linked male-specific transplantation antigen, but others were unable to confirm this (Gasser *et al.*, 1974). All other attempts to use this experimental protocol, with Y chromosomes derived from a variety of strains of mice, including wild mice of separate subspecies, have failed to find a Y-linked polymorphic transplantation antigen (Zaalberg, 1959; Andrews and Wachtel, 1979; Simpson *et al.*, 1979). Moreover, Johnson (1982) devised an ingenious protocol, using haemopoietic chimeras, to search for an H-Y_T polymorphism, even if it were not Y-linked, but he too has so far failed to find one.

In the absence of a polymorphism, concordance between the possession of a Y chromosome and the expression of H-Y_T, irrespective of phenotypic sex, would provide circumstantial evidence of Y linkage. Accordingly, several authors have attempted to correlate the ability of male-derived tumours to grow in female hosts with their loss of a Y chromosome. In one tumour subline that grew well in female hosts Hauschka and Holdridge (1962) did find fewer cells carrying a Y chromosome compared with a subline that grew poorly. Similarly, Bunker (1966) found that sublines of a retransplantable male teratocarcinoma which failed to grow in female hosts retained a Y chromosome, whereas this chromosome was lost from sublines that did grow in females. On the other hand, Eichwald and Davidson (1968) found no such correlation in six tumours that they studied. Given the complexities of susceptibility to tumour growth together with the variability in H-Y_T expression and response to it, the absence of correlation in some instances is perhaps less telling than the presence of a correlation in others.

At the organism level a number of sex chromosome abnormalities exist in the laboratory mouse and provide evidence of Y-chromosome association of the H-Y_T antigen. I have already mentioned the *tfm*, XY females in this regard. Other experiments have demonstrated that XO mice (phenotypic females) lack H-Y_T antigen whereas XXY mice (phenotypic males) express H-Y_T antigen (Celada and Welshons, 1963; Simpson *et al.*, 1982). Thus expression of H-Y_T is not merely due to the presence of a single X chromosome and, likewise, its absence is not due to the presence of two X chromosomes. By contrast an apparent exception to the concordance of H-Y_T and possession of a Y chromosome was the finding that *Sxr*, XX sex-reversed mice, which are phenotypic, though sterile, males with testes (Cattanach *et al.*, 1971), do express H-Y_T antigen (Bennett *et al.*, 1977). However, it has been recently shown (Singh and Jones, 1982; Evans *et al.*, 1982: McLaren and Monk, 1982) that *Sxr* results from an inversion of the Y chromosome so that the normally pericentric region, thought to contain the male sex-determining locus and putative H-Y_T locus is relocated to the teleomeric region where it can undergo meiotic crossing-over with the X chromosome. In about 50% of the progeny that derive their paternal X chromosome from an *Sxr*, XY father, the X chromosome carries a piece of the inverted Y chromosome, presumably including the male-determining locus, and they develop as males. These were long thought to have an XX chromosome constitution as the translocated Y went unnoticed, and the 50% segregation ratio gave the false impression that *Sxr* was an autosomal, dominant mutation.

In the examples quoted so far the expression of H-Y_T has correlated with presence of testes, although not necessarily with secondary male characteristics. Recently it has been found that two Y chromosomes derived from wild populations of *Mus domesticus*, when introduced into certain laboratory strains of mice, lead to the development of XY phenotypic females with ovaries (Eicher *et al.*, 1982; Eicher, 1982). In both cases the XY females were found to express H-Y_T antigen by skin grafting (Simpson *et al.*, 1983), again indicating that the expression of H-Y_T seems to be associated with the presence of a Y chromosome, rather than with phenotypic or gonadal sex.

7.2.3 H-Y_T in other species

The direct demonstration of a sex-linked transplantation antigen is difficult in the absence of inbred strains because its effects are rapidly obscured by the background of many other major and minor allogeneic transplantation antigens. Thus most direct evidence for an H-Y_T antigen in other species has come from inbred strains of laboratory animals, especially rats (Billingham and Silvers, 1959; Billingham *et al.*, 1962; Zeiss *et al.*, 1962; Chen and Silvers, 1982) and rabbits (Chai, 1968), which have both been shown to exhibit a male transplantation antigen similar to the murine H-Y_T. Moreover, rat H-Y_T was inferred to be antigenically identical to murine H-Y_T since the inoculation of rat cells into neonatal female mice can render them tolerant of subsequent grafts of male mouse skin (Silvers and Yang, 1973). Although only some strains of rat produce this effect, suggesting to some the existence of strain differences in H-Y (Hildemann, 1974), it was thought by others to be a technical problem due to the more rapid destruction by mice of rat cells from some strains than from others (Gasser *et al.*, 1974). This is a general problem, which makes it difficult to use this technique to establish the similarity, or otherwise, between the H-Y_T antigens of more disparate species. However, in support of the argument that H-Y_T is polymorphic in rats, Mullen and Hildemann (1972) reported that in some strain combinations F_1 hybrid males rejected skin grafts from males of the maternal parental strain, indicating Y linkage.

I have already mentioned evidence suggesting the existence of an H-Y_T-like antigen in man. More wide-ranging studies have also indicated the existence of sex-linked transplantation antigens in species belonging to other classes. Thus Miller (1962) observed a male-specific transplantation antigen in inbred strains of platyfish. Birds, too, appear to possess a sex-linked transplantation antigen, but it is expressed by the females which, in this group, constitute the heterogametic sex. Following the early observations of Kozelka (1932) several authors have used the more recently developed inbred strains of chickens to show that female-to-male grafts are often rejected whereas grafts in other sex combinations survive (Bacon and Craig, 1966, 1969; Bacon, 1970; Gilmour, 1967). As in the mouse and rat there appear to be rejector and non-rejector strains, and in one limited study using the parental male to F_1 male graft technique, McCarrey *et al.* (1981) failed to find evidence of a polymorphism. Since the sex chromosomes of birds have been designated Z and W with females being ZW and males ZZ (Miller, 1938), to distinguish them from the mammalian X and Y system, the female transplantation antigen of birds is often referred to as H-W; to be consistent with the nomenclature adopted in this chapter I shall refer to it as H-W_T. Thus, sex-limited transplantation antigens, possibly identical from the studies in mouse and rat, seem to be found throughout phylogeny.

7.3 H-Y (H-Y_C) DETECTION BY CELL-MEDIATED CYTOTOXICITY *IN VITRO*

7.3.1 The female cytotoxic T-cell response to male cells

Lymphocytes obtained from some strains of female mice which have been challenged with a syngeneic male skin graft often exhibit an ability to specifically kill syngeneic male but not female cells (lymphocytes are usually used for convenience) in culture (Goldberg *et al.*, 1973). As in skin graft rejection, no response of male mice challenged with syngeneic female cells can be detected. I shall refer to the antigen detected as H-Y_C. Gordon *et al.* (1975) showed that the effector cells in this system are cytotoxic T lymphocytes (CTL) which only kill male targets carrying an H-2 class I antigen in common with themselves. Extensive studies using F_1 hybrid responder females revealed that the anti-(H-Y_C) CTLs generated usually show a preferential specificity for killing male target cells that share with them a particular H-2 antigen, coded by either the *H-2K* or *H-2D* subregion of the MHC depending upon the particular *H-2* haplotypes involved (Hurme *et al.*, 1977). For example, CTLs derived by immunizing C57BL/10 ($H\text{-}2^b$) females with C57BL/10 males kill any male, but not female, target cells which also bear the $H\text{-}2D^b$ antigen, whereas CTLs derived by immunizing (B10.D2×C57BL/10)F_1 hybrid females ($H\text{-}2^d/H\text{-}2^b$) with B10.D2 males ($H\text{-}2^d$), only kill male target cells which also bear the $H\text{-}2K^d$ antigen. This H-2 restriction of CTLs is not unique to the $H\text{-}Y_C$ target antigen but is also seen in connection with the response to virally infected cells (Zinkernagel and Doherty, 1974) and to other minor histocompatibility antigens (Bevan, 1975). It is thought that there is some kind of associative response of the CTLs to both H-2 and the target antigen; either T lymphocytes may need to recognize both self H-2 and the target antigen ($H\text{-}Y_C$) before they can function (dual-recognition hypothesis) or the H-2 and target ($H\text{-}Y_C$) molecules may interact to generate a new antigenic specificity which the lymphocytes 'see' (altered-self) (Zinkernagel and Doherty, 1974).

The H-2 restriction of the CTLs does not imply that the immune response is to a polymorphic male-specific antigen encoded by an MHC-linked gene. As in the case of the skin graft assay, females of one strain (e.g. C57BL/10, $H\text{-}2^b$) can be presensitized equally well for the subsequent production of male-specific CTLs by immunization with male cells from a syngeneic, an H-2 congenic (e.g. B10.BR, $H\text{-}2^k$) or an allogeneic (e.g. CBA, $H\text{-}2^k$) strain of mouse, irrespective of its H-2 type, and whether or not male-specific CTLs can be generated in females of that strain (Gordon *et al.*, 1976; Simpson and Gordon, 1977). Further evidence that it is the H-2K/D antigens themselves which are involved directly in CTL restriction, comes from data showing that anti-($H\text{-}2K^k$) monoclonal antibodies can block the killing of $H\text{-}2K^k$ male target cells by $H\text{-}2K^k$-restricted anti-($H\text{-}Y_C$) CTLs (Fisher-Lindahl and Lemke, 1979). Similarly anti-($H\text{-}2D^b$) serum can block killing by $H\text{-}2^b$-restricted anti-($H\text{-}Y_C$) CTLs (Koo *et al.*, 1979). The blocking results from binding of the antibody to the target cells, not to the CTLs.

As in the rejection of male skin grafts, the MHC plays an important role in determining whether a female mouse will produce anti-($H\text{-}Y_C$) CTLs in response to a challenge of syngeneic male cells (Gordon *et al.*, 1975), although non-MHC-linked factors may also be important (Fierz *et al.*, 1982). Females of many strains of mice fail to produce anti-($H\text{-}Y_C$) CTLs, but females carrying the $H\text{-}2^b$ haplotype always respond in this way (Gordon *et al.*, 1975). However, in contrast to the IB^b allele required for skin graft rejection, the dominant gene of the $H\text{-}2^b$ haplotype controlling CTL response maps to the *IA* subregion (Hurme *et al.*, 1978b). Also, F_1 hybrids between two non-responder strains may be responders, apparently because of complementation between two genes which map to the *IC* subregion of the MHC (Hurme *et al.*, 1977, 1978b; see also Simpson, 1982b for detailed discussion). In summary, it appears that only certain *I*-region alleles permit T-cell help for the expansion of CTLs which then recognize H-Y_C antigen in the context of the H-2K or D antigens expressed on the target cells.

7.3.2 The H-Y antigens detected by transplantation and CTL assays are probably the same

It is evident from these studies that the lack of response to H-Y_C in some strains is not due to males lacking the antigen. Indeed the allogeneic stimulation experiments indicate that the antigenic determinant is the same in all strains, irrespective of their ability to respond. Again, as in the case of the male transplantation antigen, H-Y_T, no evidence of a Y-chromosome-linked polymorphic variant of H-Y_C has been obtained (Simpson *et al.*, 1979). However, like H-Y_T, H-Y_C expression correlates with the presence of a Y chromosome or Y-chromosome-derived element irrespective of phenotypic sex. Thus XO females lack H-Y_C (Simpson *et al.*, 1982) whereas *Sxr*, XX males express H-Y_C (Simpson *et al.*, 1981). Also, the XY phenotypic females with ovaries, produced when the two Y chromosomes isolated from wild populations were introduced into the C57BL/6 genetic background, expressed H-Y_C (Simpson *et al.*, 1983). In all of these experiments a correlation has always been found between the detection of a male-specific antigen by the CTL assay *in vitro*, and by the transplantation assay *in vivo*. However, whereas the *ability* to reject a male skin graft depends upon a locus within the *IB* subregion of the MHC, the ability to generate male-specific CTLs is regulated by loci within the *IA* and *IC* subregions (Hurme *et al.*, 1978a). This could imply that two distinct male-specific antigens are being assayed, but in light of the complete concordance of male antigen expression detected by the two techniques it is thought more likely that the discrepancies in responsiveness reflect the involvement of different subsets of T-lymphocyte effector cells and so is a function of the complexity of the immune system, not of the antigen to which it is responding (Liew and Simpson, 1980; Greene *et al.*, 1980; Simpson *et al.*, 1982).

7.4 SEROLOGICALLY DEFINED H-Y ANTIGEN (H-Y_S)

7.4.1 Serological detection of H-Y

Goldberg *et al.* (1971) reported that serum from females who had rejected several syngeneic male skin grafts was cytotoxic for mouse sperm in the presence of complement. This activity could be adsorbed by male cells but not by female cells, indicating that it was directed to a male-specific antigen. Interestingly, Bennett and Boyse (1973) found that artificial insemination of mice with sperm that had been treated with such sera and complement resulted in a small deficiency of male offspring (45.4% compared with 53.4% in controls) suggesting that Y-bearing sperm had been preferentially killed. It may also be mentioned here that these sera are cytotoxic for about 50% of pre-implantation mouse embryos (Krco and Goldberg, 1976); Epstein *et al.* (1980) subsequently showed that, indeed, it is the male embryos which are killed while the females survive. This male antigen detected by female anti-(syngeneic-male) sera was found in all strains of mice tested, and was considered to be the H-Y male antigen detected by transplantation assays. Here I shall refer to it as H-Y_S. Further results of Goldberg *et al.* (1972) indicated no correlation between the ability of female mice to reject skin grafts and their formation of antibodies to H-Y_S: non-rejector as well as rejector females produced male-specific antisera when grafted with male skin, implying that these antibodies do not play an important role in the graft-rejection process.

Unfortunately the male-specific antisera produced by skin grafting, or more conveniently by multiple injections of male spleen cells, are uniformly weak, with titres of 1:4 to 1:8 at best, so that only small amounts of well-characterized mouse anti-(H-Y_S) sera are obtainable. To compound this problem, it is notoriously difficult to distinguish between live and dead sperm in cytotoxicity assays, and these anti-(H-Y_S) sera are rarely able to kill specifically more than 25–30% of the target sperm. Other, potentially more convenient target cells, such as thymocytes or splenocytes, are not susceptible to lysis by these sera, whereas tail epidermal cells, which are killed in a sex-specific manner (Scheid *et al.*, 1972), are inconvenient to obtain. Nevertheless, despite these difficulties, sex-specific adsorption of the cytotoxic activity from female anti-(syngeneic-male) sera, tested on sperm or epidermal target cells, has become the primary serological test for H-Y_S antigen (Wachtel, 1977). Numerous variants of this test have also been employed (see Koo, 1981) in efforts to obtain a simpler assay. These include the use of rat female anti-(syngeneic-male) sera, tested on rat epidermal cells, and the use of other indicators of antibody–antigen interaction. Space does not permit a detailed description of all of these variant assays here, but the diversity of tests should be remembered when conflicting results concerning H-Y_S are reported. In particular, two other assays deserve some attention: Fellous *et al.* (1978) found that sera from female rats immunized with syngeneic male rat spleen cells were cytotoxic for the human male lymphoid cell line RAJI. Since this activity could be specifically adsorbed by male but not female cells of mice (and other mammalian species, see below) it was proposed that this assay (the 'RAJI assay') also detects the H-Y_S antigen. Because of its convenience it has been adopted by several groups. However, notwithstanding the male-specificity of the antigen detected, there has been no definitive proof that it recognizes the same antigen as those assays which employ sera and target cells of the same species. Also, because it employs rodent sera tested on human target cells it is essential to rigorously adsorb out any heterophile antibodies that may be present, but in many reports it is evident that such antibodies have not been completely removed from the sera.

A second type of assay which should be mentioned here is one which uses a monoclonal anti-(H-Y) antibody produced by a hybrid myeloma cell line. The details of one such anti-(H-Y) monoclonal antibody have been reported (Koo *et al.*, 1981). In principle, the production of H-Y-specific monoclonal antibodies should obviate the problems related to the scarcity of well-defined anti-(H-Y_S) sera, but so far the progress in this direction has been disappointing (see Simpson, 1982a) and most of the available data on H-Y_S expression have been obtained using the conventional antisera assays.

Unlike the transplantation and cytotoxic T-cell assays, the specific adsorption of anti-(H-Y) activity from sera provides a direct method for examining whether a similar antigen is expressed by species other than mice. Wachtel *et al.* (1974) used this approach to show that not only could a cross-reacting H-Y_S antigen be detected in various rodents (rats guinea pigs) and related species (rabbits), but that humans also expressed an antigen which would adsorb anti-(mouse H-Y_S) activity. Moreover, in all mammalian species this antigen was also found to be expressed by males but not by females. In a further study (Wachtel *et al.*, 1975b), it was found that a cross-reactive H-Y_S antigen could also be detected in birds (chickens) and in amphibia. Again, only one sex expressed H-Y_S, but this was the heterogametic sex in each case. Thus female, but not male, chickens expressed H-Y_S, correlating with the observation that male chickens reject female grafts, but not vice versa. Among the amphibia tested, in *Rana pipiens* it was the males which expressed H-Y_S, whereas in *Xenopus laevis* it was the females. Sex-specific H-Y_S antigen was subsequently reported in turtles (Engel *et al.*, 1981a; Zaborksi *et al.*, 1982), various species of fish (Pechan *et al.*, 1979; Shalev and Heubner, 1980; Müller and Wolf, 1979; Shalev *et al.*, 1978; Wiberg, 1982) and even in lobsters and two species of insects (Shalev *et al.*, 1980a). When the data were available, H-Y_S expression was generally found to be associated with heterogamy, although in the two insect species H-Y_S was only detected in the homogametic females. The significance of this is not entirely clear since many of these non-mammalian species may be induced to develop as fully fertile males or females irrespective of their sex chromosome constitution.

Several attempts have been made to use the anti-(H-Y_S) sera to biochemically identify the molecule carrying the H-Y_S antigenic determinant. Thus Hall and Wachtel (1980) reported specific immunoprecipitation of polypeptides with molecular weights of 18000 and 31000 from lysates of Sertoli cells surface-labelled with ^{125}I. Another biochemical approach to identifying the H-Y_S antigen followed from the finding of Fellous *et al.* (1978) and Beutler *et al.* (1978) that cell surface expression of H-Y_S antigen appeared to require its association with β_2-microglobulin (β_2m). Daudi, a human male lymphoid cell line lacking the capacity to express β_2m owing to mutation, appears to secrete H-Y_S into the culture medium (Ohno *et al.*, 1979). This form of H-Y_S was reported to be associated with a polypeptide of 15000–18000 molecular weight

(Ohno *et al.*, 1979; Nagai *et al.*, 1979; Hall and Wachtel, 1980; Hall *et al.*, 1981). Finally, following the suggestion that H-Y antigen may be actively secreted by male gonadal cells (Ohno *et al.*, 1976), Müller *et al.* (1978a,b) reported that material present in epidydimal fluid and in culture medium from testicular, but not non-testicular, tissues incubated *in vitro* could inhibit anti-(H-Y_S) sera. Zenzes *et al.* (1978a) showed that this putative, soluble H-Y_S antigen was secreted by the Sertoli cells, but specific precipitation by anti-(H-Y_S) sera of radiolabelled proteins from these culture media has not been achieved (Hall and Wachtel, 1980; Gore-Langton *et al.*, 1983).

Whether or not the association of the H-Y_S antigen with a polypeptide(s) is eventually confirmed, its phylogenetic conservation provokes the thought that the antigenic determinant might be a carbohydrate. In support of this idea, Shapiro and Erickson (1981) showed that H-Y_S antigen on the surface of cells was susceptible to degradation by periodate and by β-galactosidase and galactose oxidase whereas it was insensitive to several proteolytic enzymes. If H-Y_S is indeed a carbohydrate antigenic determinant, it might well be found associated with various polypeptides and even glycolipids.

7.4.2 Genetics of H-Y_S expression

As in the case of the H-Y antigen(s) defined by the other assays, no evidence of a polymorphic variant of the serologically defined H-Y_S antigen has been forthcoming. Consequently all attempts to ascertain whether it is encoded by the Y-chromosome have been equally indirect. However, the expression of an apparently identical H-Y_S antigen in species other than the mouse, and the presumption that the linkage groups associated with the mammalian sex chromosomes have remained unchanged throughout phylogeny, allow conclusions to be based on a wider variety of data. Nevertheless, no firm conclusions are yet possible, although like H-Y_T and H-Y_C, it is evident that expression of H-Y_S is not dependent upon the action of the androgenic hormones, at least by their usual mechanism of inducing male secondary sexual characteristics. Thus both humans and mice with *testicular feminization* syndrome (i.e. genotypically *tfm*, X/Y) were found to express H-Y_S (Bennett *et al.*, 1975; Koo *et al.*, 1977b).

Wachtel *et al.* (1975c) reported that lymphocytes from human males with two Y chromosomes (XXYY or XYY) adsorbed more H-Y_S activity from mouse anti-(H-Y_S) sera than did lymphocytes from normal males with a single Y chromosome. The implied increased level of H-Y_S in the males with two Ys would be most readily explained by an increased dosage of H-Y_S genes, suggesting that H-Y_S is encoded by the Y chromosome. Unfortunately no true quantitative adsorptions were made, presumably because of the paucity of reagents, and so the data obtained were only semiquantitative and must be regarded as provisional. Other human subjects with rearrangements of the Y chromosome have also been studied in order to attempt to assign the putative Y-linked H-Y gene to a particular region of the Y chromosome. Thus two people who were phenotypically females, but with many of the characteristics of Turner's syndrome, and who were typed H-Y_S-negative, were found to have a 46,X i(Yq) karyotype (Koo *et al.*, 1977a; Rary *et al.*, 1979). That is, they lacked the short arm of the Y chromosome (Yp) implying that H-Y_S is specified by a gene located in this chromosomal region. In another case of a phenotypic female with a 46,XYp^- karyotype (i.e. the short arm of the Y was deleted), Rosenfeld *et al.* (1979) were unable to conclude whether there was a complete absence of H-Y_S or merely a reduced level. In concert with a suggestion that there may be a cluster of multiple genes encoding the H-Y_S antigen (Bennett *et al.*, 1977; Wachtel and Ohno, 1979), they proposed that, if the cluster were located in the pericentromeric region of the Y chromosome (Koo *et al.*, 1977a), only a fraction of the multiple H-Y_S loci had been deleted in this Yp^- individual. Koo *et al.* (1977) have also found an individual lacking the short arm of the Y chromosome and definitely expressing H-Y_S. They suggested that frequent pericentric inversion may result in the H-Y_S locus being situated in the centromeric region of Yq or Yp in different people.

By contrast, other data have been presented indicating H-Y_S expression in the absence of any detectable Y-chromosomal component, in XO female mice and humans, and in human females with Turner's syndrome, gonadal dysgenesis and X-chromosome rearrangements, particularly those in which the short arm of one X is deleted (Wolf *et al.*, 1980a,b; Engel *et al.*, 1981b; Wachtel *et al.*, 1980a; Haseltine *et al.*, 1982; Koo *et al.*, 1981). Other reports have indicated H-Y_S expression by trans-sexual humans with an apparently normal XX karyotype and ovaries (Engel *et al.*, 1980; Špoljar *et al.*, 1981). The reagents used in these various studies by different authors have covered the range of H-Y_S tests commonly employed and included the use of mouse female anti-(syngeneic-male) sera, tested on mouse sperm, rat female anti-(syngeneic-male) sera tested on rat epidermal cells, the 'RAJI assay', and a monoclonal antibody, thus implying general agreement as to the conclusion that H-Y_S may be expressed in the absence of Y-chromosomal material. Consequently it has been proposed that the H-Y_S structural gene is located on the X chromosome or on an autosome, that its expression is normally repressed by the activity of an X-chromosomal gene (located on the short arm of the X in humans) and that it is activated by the product of a Y-chromosomal regulator gene (Wolf *et al.*, 1980a). The H-Y_S gene(s) mapped to the short arm of the Y-chromosome by Koo *et al.* (1977a) would thus be regulatory rather than structural (Wachtel *et al.*, 1980a). In regard to this hypothesis the cases of sex-reversed XY human females with campomelic dysplasia, thought to be a recessive genetically determined mesenchymal disease, are of some interest. Evidently some, but not all, of these individuals are H-Y_S-negative (Bricarelli *et al.*, 1981; Puck *et al.*, 1981) implying that the autosomal or X-linked gene causing this condition can also suppress H-Y_S expression. Other cases of XY females lacking H-Y_S antigen in both wood lemmings (Wachtel *et al.*, 1976b) and in humans (Bernstein *et al.*, 1980) have also been explained by X-linked regulatory loci. Nevertheless, it is difficult to understand, by this hypothesis, why some human females with Turner's syndrome, and lacking the short arm of the Y chromosome although possessing part of the long arm, should be H-Y_S-negative, whereas phenotypically similar individuals lacking the complete Y chromosome should be H-Y_S-positive.

The various instances of XX males, of which the *Sxr* (*sex-reversed*) mutation in the mouse is the prototype, are relevant to this discussion, although they can be explained equally well by hypotheses which locate the H-Y_S structural genes on the Y chromosome, or elsewhere in the genome; any hypothesis, however, must accord at least a regulatory function to the Y chromosome. The fact that *Sxr,* XX male mice, expressing H-Y_T and H-Y_C antigen(s), arise because of the transfer of a section of the Y chromosome to the X, has already been discussed. In accord with these data, these mice also express the serologically detected H-Y_S antigen (Bennett *et al.*, 1977). Sporadic and familial cases of males, or true hermaphrodites with an XX karyotype also occur in humans and other species and they have usually been typed H-Y_S-positive (Wachtel *et al.*, 1976a). In one human pedigree (de la Chapelle *et al.*, 1978), as well as in the case of goats carrying the *Polled* mutation (Wachtel *et al.*, 1978; Shalev *et al.*, 1980b), such sex reversal is inherited as a recessive autosomal trait, in contrast to the *Sxr* mutation in mice where it is dominant in XX individuals. Not only do these XX males (both humans and goats) express H-Y_S, but so do their mothers, although to a lesser degree than normal males or their XX sons. It was suggested that these conditions result from a karyotypically undetected Y to autosome translocation of part of a postulated cluster of H-Y_S genes. Thus heterozygotes would express lower levels of H-Y_S than homozygotes or normal males because they lack a full complement of the postulated multiple H-Y_S genes (see below for discussion of H-Y_S and sex determination in this case). However, it could equally be that an X chromosome or autosome structural locus determining expression of H-Y_S antigen, normally under control of a Y-linked regulatory gene, has become constitutively expressed because of a mutation. Such an explanation would accord with the hypothesis of Wolf *et al.* (1980a).

Much of the discussion of H-Y antigen genetics tacitly assumes that the molecule is a protein and hence has a 'structural gene'. However, as already discussed, one set of data (Shapiro and Erickson, 1981) argues in favour of the antigenic determinant being a carbohydrate. This being the case, the various postulated H-Y_S loci on the X and Y chromosomes and the autosomes could encode glycosyltransferases which act successively on some acceptor molecule, polypeptide or lipid, to build an oligosaccharide with H-Y antigenic activity. Such an hypothesis has been advanced by Adinolfi *et al.* (1982). They proposed that a core oligosaccharide, or 'precursor substance' (H-Y_p), is elaborated under the control of an autosomal gene. A glycosyltransferase, encoded by the Y chromosome, was then postulated to convert H-Y_p to H-Y_a, say by the addition of a particular monosaccharide, whereas another, encoded by the X chromosome, adds a different monosaccharide to H-Y_p to produce a different end product H-Y_r. Accordingly H-Y_a would be found only in XY individuals whereas H-Y_r would be found in both XY and XX individuals. Little or no precursor H-Y_p would usually exist in either sex, but were either of the X- or Y-linked glycosyltransferases to be deleted by chromosomal rearrangements, then significant levels of H-Y_p might remain unconverted to the end products. The authors suggested that classically produced anti-(H-Y_S) sera raised in female mice recognize the male-specific H-Y_a, and may sometimes cross-react with H-Y_p. On the other hand neither female anti-male nor male anti-female sera would contain antibodies recognizing H-Y_r since this molecule is expressed in both sexes. Reactivity of female anti-male sera with cells from Turner's females is explained by presuming some cross-reactivity with the H-Y_p precursor substance which would exist at higher than normal levels in these individuals. This hypothesis could explain much of the genetic data that have been reported concerning H-Y_S, and finds a precedent in the biochemistry of the human ABH and Le blood group substances. However, it still fails to explain the contradictory observations of no H-Y_S antigen in certain females with Turner's syndrome and rearranged Y chromosomes, but its presence (H-Y_p) in other such females with no Y chromosomes. Further, the *ad hoc* assumptions concerning cross-reactivity of H-Y_a and H-Y_p, although plausible, make the hypothesis amenable to almost any variations in the data and so not readily testable. Resolution of the genetics of H-Y_S must await the development of better serological assays and the determination of its biochemical nature.

7.4.3 On the relationship of H-Y_S to H-Y_T and H-Y_C

I have already discussed the view that the male-specific antigen detected in mice by the transplantation assay is the same as that detected by the CTL assay, although different subsets of effector T cells may be involved. However, various data suggest that this antigen is distinct from the male-specific antigen detected serologically (Silvers *et al.*, 1982; Simpson *et al.*, 1982). Melvold *et al.* (1977) reported a single, phenotypically male, mutant mouse which failed to express the H-Y_T antigen; skin from this mouse was permanently accepted by both male and female syngeneic (but not mutagenized) hosts, whereas it rejected syngeneic male but not female grafts. On the other hand, H-Y_S antigen was detected serologically. Unfortunately this male was sterile and so further genetic analysis was precluded. Subsequently, XO female mice, which do not express H-Y_T or H-Y_C (Celada and Welshons, 1963; Simpson *et al.*,

1982) have been reported to express serologically detectable H-Y_S (Engel *et al.*, 1981b). Finally, there are the extensive data, discussed in the preceding section, indicating H-Y_S expression in humans in the absence of a Y chromosome, whereas whenever studied in the mouse there has been a constant correlation between the expression of H-Y_T and H-Y_C and the presence of a Y chromosome (Simpson *et al.*, 1983).

The bulk of the available genetic data supports the view that these antigens are distinct. On the other hand, Koo and Varano (1981) have found that a monoclonal antibody that reacts with H-Y_S also blocks H-Y_C-dependent cell-mediated cytotoxicity. This may imply identity, although as Simpson *et al.* (1982) pointed out, this effect only indicates close proximity of the antigens on the cell surface such that steric hindrance could occur.

Despite these arguments, the problems with the serological assays suggest that a final conclusion is not yet warranted. By the hypothesis of Adinolfi *et al.* (1982), for example, it is possible that anti-(H-Y_S) sera may contain antibodies corresponding in specificity to that of the T cells which mediate transplantation rejection and cytotoxicity, as well as antibodies of different but related specificity (e.g. to the postulated H-Y_p precursor substance). The generation of new monoclonal antibodies with anti-(H-Y_S) activity and the analysis of their reactivities with individuals of different genotypes may help to resolve the question.

7.5 H-Y ANTIGEN AND SEX DETERMINATION

7.5.1 The hypothesis

During embryogenesis in mammals, ovaries or testes differentiate from bipotential gonadal primordia according to the genotype of the embryo. All subsequent events of sex-specific differentiation (e.g. degeneration of the Müllerian or Wolfian ducts; development of external genitalia, etc.) occur in response to hormones produced by the ovaries or testes once they have differentiated. Usually, in most mammalian species, the presence of a Y chromosome is sufficient to ensure testicular differentiation, whereas in its absence ovarian differentiation occurs. Evidently the Y chromosome carries a testis-determining gene (for reviews see Ohno, 1971; Davis, 1981; Gordon and Ruddle, 1981; Short, 1979). In lower vertebrates the same situation may hold, or as in birds, for example, ovarian differentiation may occur in the presence of a W chromosome with testicular differentiation in its absence (e.g. see Bloom, 1974). However, in these species the direction of differentiation is more plastic than in mammals and may be influenced by environmental factors (e.g. temperature) or by injections of female or male hormones.

How genotypic sex is translated into phenotypic sex is a longstanding question. Arguing that the primary mechanism must be simple to avoid undue genetic load from mutation; that a cell-surface location for a molecule guiding organogenesis is likely; and that the highly conserved nature of the H-Y antigen expressed by the heterogametic sex implies a constant function related to sexual development, Wachtel *et al.* (1975a) proposed that this molecule is indeed the Y-chromosome-encoded factor responsible for testicular differentiation of the bipotential gonadal primordia in mammals. Further, since in other vertebrates H-Y (or H-W) antigen is detected in the heterogametic sex, whether male or female, it was postulated that when females are the heterogametic sex H-Y (H-W) antigen induces ovarian differentiation. A considerable amount of data bearing on this hypothesis has now been collected; some have suggested that these data fail to support the idea that the H-Y antigen, however assayed, has a role in testicular (or ovarian) differentiation (e.g. Silvers *et al.*, 1982); others argue that the hypothesis in its broadest terms has been vindicated, although details may need to be modified (e.g. see Wachtel, 1983).

7.5.2 Genetic evidence

Although in mammals, male sexual differentiation depends on the presence of a Y chromosome, the many exceptions, which may occur for a variety of reasons, provide situations in which to test the correlation between H-Y antigen expression and testicular differentiation. By the hypothesis, and given that the phylogenetic conservation of the H-Y antigen implies that the antigenic determinant itself functions in promoting testicular differentiation, testicular differentiation should not be observed in the absence of H-Y antigen. The converse would, however, not necessarily be true; if testicular differentiation, initiated by H-Y antigen, were to require a series of subsequent molecular steps for its completion, we could easily imagine the presence of H-Y antigen without testicular differentiation because of some defect in one of the subsequent steps. The deletion of a receptor for H-Y on specific target cells, assuming a hormone-like role for H-Y, has been one suggestion of how mutations could prevent its action.

In the mouse, in humans, and in goats I have already recorded several instances of testicular differentiation occurring together with expression of H-Y antigen (defined by all three groups of assays) in XY individuals but with female secondary sexual development (*testicular feminization*) and in individuals with male development but apparently lacking a Y chromosome (*Sxr* in the mouse, XX male syndrome in man and *Polled* in the goat). Human females with XY karyotype, no testicular tissue and no H-Y_S antigen have also been reported (Bricarelli *et al.*, 1981; Puck *et al.*, 1981; Bernstein *et al.*, 1980), while among other species wild populations of the wood lemming exhibit a polymorphism for an X-linked mutation which permits XY individuals to develop as fertile females. In one study these were typed as H-Y_S-negative (Wachtel *et al.*, 1976b), although Wiberg *et al.* (1982) has reported that they express H-Y_S. Even though these contradictory results are disturbing in connection with the accuracy of H-Y_S serology, either result could be compatible with the normal involvement of H-Y antigen in testicular differentiation. The same is true of the apparent H-Y_S expression observed in Turner's syndrome females. It is conceivable that testicular differentiation is prevented by such karyotypic abnormalities by a mechanism which acts in the sequence of differentiation subsequent to the point of action of H-Y_S. Also in certain cases of evidently fertile females expressing H-Y_S, such as in the mothers of the XX men studied by de la Chapelle *et al.* (1978) and in female goats heterozygous for the *Polled* gene (only the homozygotes are sex-reversed in this case) (Wachtel *et al.*, 1978), the level of expression of H-Y_S was reported to be lower than in normal males; on the basis of this and other data it was proposed that there is a threshold effect such that testicular differentiation depends upon the expression of a critical amount of H-Y antigen (Wachtel and Ohno, 1979).

Although these data may provide information concerning the putative mechanism of induction of testicular differentiation by H-Y antigen, they clearly provide no crucial test of the hypothesis. Results which would tend to disprove the hypothesis would be the observation of testicular differentiation in the absence of H-Y antigen. Indeed several such instances have been reported. The mutant male mouse of Melvold *et al.* (1977) clearly lacked H-Y_T, detected by skin grafting, suggesting that H-Y_T is not male-determining; H-Y_S was, however, detected. Nevertheless this mouse was sterile and its testes were histologically abnormal. Subsequently, three reports have been published, describing several human males (actually pseudohermaphrodites and male to female trans-sexuals), with a 46,XY karyotype and testicular tissue, but lacking H-Y_S (Haseltine *et al.*, 1981; Špoljar *et al.*, 1981; Engel *et al.*, 1980). It was suggested that in these subjects the H-Y_S molecule had lost its antigenic activity while retaining its functional activity. This produces a dilemma since it would imply that the antigenic determinant itself does not function in regulating testicular differentiation; however, it was on the basis of its evident phylogenetic conservation, that the H-Y_S antigenic determinant was proposed to have a highly conserved function – namely, to promote testicular differentiation of the gonad. If the antigenic determinant does not have this function, then it is necessary to look elsewhere for a reason for its conservation, as opposed to the possible conservation of the molecule carrying this determinant. However, it cannot be excluded that a small number of testis cells, sufficient to cause testicular development, did express H-Y_S in these individuals, although in several of the cases testicular biopsies were typed H-Y_S-negative. Alternatively it could be that H-Y_S was expressed only early on in embryonic development, contrary to its usually ubiquitous expression. Thus, although these data are unfavourable to the general hypothesis, they provide no definitive rebuttal.

With regard to the possible conservation of the Y-linked molecule that induces testicular differentiation, the recent observations of Eicher and her colleagues (Eicher, 1982; Eicher *et al.*, 1982) on the introduction of Y-chromosomes from wild mouse populations into the C57BL/6 inbred laboratory strain are of some interest. These Y chromosomes, in the C57BL/6 background, are deficient in their ability to cause male development; XY individuals develop as females with ovaries although the same Y chromosomes are evidently able to cause normal male development in the stock from which they were derived. This implies that male development results from an interaction of a Y-linked gene and X- or autosome-linked genes; since different Y chromosomes on the same background produce different effects (the C57BL/6 Y chromosome causes normal male development) it seems likely that the male-determining gene of the Y chromosome is *not* phylogenetically conserved but is, indeed, polymorphic. The XY females produced in these stocks have been shown to be $H\text{-}Y_T$- and $H\text{-}Y_C$-positive (Simpson *et al.*, 1983); they have not yet been tested for $H\text{-}Y_S$ expression.

7.5.3 Evidence from cell biology

Two observations have permitted a direct approach to testing whether $H\text{-}Y_S$ antigen is responsible for testicular differentiation in mammals, and ovarian differentiation in birds. When twins occur in cattle, chorionic vascular anastomoses result in haemopoietic chimaerism; further if the twins are of opposite sex, the female twin is virilized (the freemartin condition), its gonads developing as testes. On the basis of the frequent lack of detectable cellular chimaerism but the presence of $H\text{-}Y_S$ antigen in the gonads of freemartins, Ohno *et al.* (1976) suggested that blood-borne $H\text{-}Y_S$ antigen, released in soluble form from the male twin, was specifically adsorbed by the female twins' gonadal primordia which were then induced to differentiate as testes. Such apparently non-cell autonomous development of genetically female gonodal cells into testicular structures in the presence of small numbers of male cells has been commonly recorded in other situations, such as the excessive development of phenotypic males amongst XY $\longleftrightarrow$ XX mouse chimaeras (see McLaren, 1976). Accordingly, the existence of a testicular-inducing substance is implied and the data of Ohno *et al.* (1976) equate this with the $H\text{-}Y_S$ antigen. The second observation is that when testicular and ovarian cells are disaggregated and then allowed to reaggregate in tissue culture, they re-form either testicular- or ovarian-like structures according to their origin. Many experiments have been made in an attempt to interfere with this reorganization by adding anti-($H\text{-}Y_S$) sera or soluble $H\text{-}Y_S$ antigen to the cultures.

Ohno *et al.* (1978) reported that when dissociated testicular cells from newborn mice were incubated in rotation culture together with normal mouse serum they reorganized into long tubular structures, whereas in the presence of anti-(H-Y_S) sera, few tubules but many follicular-like structures were formed. Similarly, Zenzes *et al.* (1978b) found that testicular-like reorganization of newborn rat testes was inhibited by anti-(H-Y_S) but not by antisera directed to MHC antigens. These experiments may be criticized since they do not clearly demonstrate that it is the anti-(H-Y_S) antibodies, rather than other components of the sera, which are effective (e.g. comparisons of the effect of sera adsorbed by male and female cells were not made). Moreover, the inhibition of a cell function by the binding of an antibody to a cell-surface ligand does not necessarily imply the specific involvement of that ligand in the cell function which is inhibited; the possibility of non-specific effects, such as steric hindrance, have to be carefully excluded. The use of anti-(MHC antigen) sera as a control by Zenzes *et al.* (1978b) is helpful but not conclusive; they found that tubular patterns of reaggregation were indeed inhibited to some extent non-specifically by these sera, although they stated that none of the follicular structures formed in the presence of anti-(H-Y_S) sera were formed in the presence of anti-(MHC antigen) sera. In another study, Müller and Urban (1981) failed to find any specific effect of anti-(H-Y_S) sera on the reaggregation of newborn rat testicular cells.

Alternative experiments of this type are to attempt to force ovarian cells to form testicular structures in culture by exposure to soluble H-Y_S antigen, obtained from Daudi cells (Nagai *et al.*, 1979), from testicular cells incubated *in vitro* (Müller *et al.*, 1978b; Zenzes *et al.*, 1978a) or from the serum of foetal bulls (Wachtel *et al.*, 1980b). It was shown that ovarian cells exposed to a source of H-Y_S antigen do, in fact, become H-Y_S-positive as assayed by their ability to adsorb specific activity from anti-(H-Y_S) sera (Wachtel *et al.*, 1980b; Müller *et al.*, 1978a, 1979a; Müller *et al.*, 1978a). Accordingly Zenzes *et al.* (1978c) reported that newborn rat ovary cells, exposed to medium from cultures of rat testicular cells, would reorganize into testicular structures; in a note added in proof they recorded that this effect was inhibited if anti-(H-Y_S) sera were added. These results were confirmed by Müller and Urban (1981). Müller *et al.* (1978c) further found that the morphological reorganization of testicular structures was accompanied by the appearance of luteinizing hormone/human chorionic gonadotropin (LH/HCG) receptors which characterize newborn rat testes but not newborn rat ovaries (Siebers *et al.*, 1977). However, Benhaim *et al.* (1982), using Daudi and teratocarcinoma culture supernatants as their source of H-Y_S, found no evidence of induction of other markers of male gonad differentiation (viz. testosterone and Müllerian-inhibiting substance) in foetal rat or calf ovaries although some of the morphological changes seen by others were noted.

Several experiments have also been conducted to examine directly the hypothesis that H-Y_S antigen may be involved in ovary formation in species in which the female is heterogametic. Male chicken embryos (ZZ), feminized by injections of oestrogen, develop ovaries which express H-Y_S (or H-W_S)

antigen, as do the ovaries of normal females (Müller *et al.*, 1979b). In a further study of quails, similarly feminized, it was shown that H-Y_S (H-W_S) induction was confined to the gonads, and no induction occurred in non-gonadal tissues (Müller *et al.*, 1980). Similarly *Xenopus laevis* males (ZZ) feminized by exposure to oestrogens, developed ovaries which were H-Y_S (H-W_S)-positive, their somatic tissues remaining H-Y_S (H-W_S)-negative (Wachtel *et al.*, 1980c). In support of the possibility that H-Y_S (H-W_S) causes ovarian organization in these species, Zenzes *et al.* (1980) demonstrated that dissociated chicken testicular cells reorganize as an ovotestis when exposed to culture supernatants of Daudi cells or rat testicular Sertoli cells. This effect was reportedly inhibited by addition of anti-(H-Y_S) sera to the incubation.

7.6 CONCLUDING REMARKS

There can be no doubt that the existence of an antigen, or antigens, specific to the heterogametic sex is widespread, at least throughout the vertebrates. Whether these sex-limited antigens, assayed by different methods, are identical, related or have no relationship to one another is still an open question. As far as the transplantation H-Y_T antigen is concerned, it appears that the antigens of rats and mice are similar if not identical; unfortunately the available experiments do not permit a test of identity between more distant species, such as mice and chickens. Evidence has been reviewed here indicating that H-Y_T and the male antigen detected by cytotoxic T lymphocytes (H-Y_C) are probably one and the same, the involvement of different sets of T lymphocytes in these assays probably reflecting complexities of the immune system, not the antigen; nevertheless the evidence is circumstantial and not conclusive. Other evidence suggests that the male-specific antigen detected by serology (H-Y_S) is distinct and phylogenetically conserved. The expression of the former, H-Y_T/H-Y_C, antigen(s) closely correlates with the presence of a Y chromosome, whereas the expression of the H-Y_S antigen does not. Whether this distinction is an artifact, perhaps stemming from the technical difficulties of the serological assays used with the attendant possibility of some incorrect typing, or whether it is real, although possibly reconcilable by the hypothesis of Adinolfi *et al.* (1982) (i.e. that H-Y is a complex carbohydrate antigen, constructed by glycosyltransferases encoded by X, Y and autosomal loci), remains to be resolved.

The question of whether any of the H-Y antigens may be involved in differentiation of the testis in mammals, or ovary in other species, also remains. For contrasting views, readers may consult Silvers *et al.* (1982) and Simpson *et al.* (1983) or Wachtel (1983). Several points, however, may be made here. The genetic data are generally inconclusive since the hypothesis that H-Y antigen is involved in testicular differentiation can be adapted to almost all the instances of abnormal sexual differentiation. On the other hand, the results obtained with reorganization of ovarian and testicular cells in culture, in response to anti-(H-Y_S) and exogenously supplied H-Y_S antigen, provide encouraging support for the hypothesis. However, in many experiments technical difficulties have resulted in less than rigorous demonstrations that either anti-(H-Y_S) antibodies or soluble H-Y_S antigen have, themselves, mediated the effects observed. For example, the finding that anti-(H-Y_S) sera, added to a crude preparation of H-Y_S antigen, inhibits the apparent ability of the antigen to cause ovarian cells to form testicular structures is not sufficient. It must be shown that other control antisera, preferably anti-(H-Y_S) sera adsorbed with male but not female cells, do not have this effect. Further it is troubling that most of the evidence is morphological in nature; in one experiment (Benhaim *et al.*, 1982) no biochemical correlates of testicular differentiation were observed when rat and cow ovarian cells were treated with H-Y_S antigen. The observation of Müller *et al.* (1978c) that LH/HCG receptors were induced in similar experiments is less than convincing since these receptors appear on ovarian cells in any case albeit somewhat later in development (Siebers *et al.*, 1977).

It is clear that the original reasoning which led to the hypothesis (Wachtel *et al.*, 1975a; Ohno, 1976) that H-Y antigen is involved in gonadal differentiation needs revising in certain aspects. Whereas it was proposed that male sexual development depends primarily on a Y-linked gene, which is likely to be highly conserved, the results of Eicher and her colleagues (Eicher *et al.*, 1982; Eicher, 1982) suggest that not only are several factors, as well as one encoded by the Y chromosome, involved but also that the Y chromosome male-determining gene is polymorphic, at least in mice. Hence the apparently highly conserved nature of the H-Y_S antigen is no longer an argument. Also, if it is not Y-linked, then it cannot be the 'primary' testis-inducing molecule. Further, the apparent existence of H-Y_S-negative individuals with testicular differentiation, if confirmed, would imply that the antigenic determinant itself, irrespective of the possible involvement of the carrier molecule, is unnecessary for testicular differentiation.

In regard to the conservation argument, it is curious that the same molecule should, in different species, promote differentiation of an ovary or a testis. Nevertheless, independent of H-Y, data have been presented suggesting that a diffusible factor from mammalian testes can cause avain testes to develop as ovaries (Weniger, 1963; Akram and Weniger, 1968). This is consistent with the H-Y hypothesis, but it was reported recently (Weniger *et al.*, 1981) that the active molecule has a molecular weight much lower than the 18000 molecular weight of the H-Y_S-reactive polypeptides identified by Ohno *et al.* (1979) and Hall *et al.* (1981); it was suggested but not proved, that testosterone is the active agent.

Finally a question which must be adequately answered is why H-Y antigen should be expressed on all tissues and not confined exclusively to the (foetal) testis. To this author it seems insufficient to attribute this to economy of control. Evidently the bipotential gonadal primordia differentiate as such before they become determined to develop further into ovaries or testes. If other factors common to both ovary and testis (e.g. the receptor for soluble H-Y_S antigen) are confined to these organs, why not also H-Y_S itself? If its sole function is to regulate development of the testis (or ovary) it might be thought that economy of energy expenditure could have led to its expression being regulated by whatever mechanisms activate other gonadal-specific functions. Such arguments are far from conclusive, but perhaps an extragonadal function should be sought.

None of the foregoing observations necessarily constitute proof for or against the hypothesis that the H-Y antigen(s), however defined, are intimately involved in gonadal differentiation. It is, however, necessary to consider them in any model of gonadal differentiation invoking H-Y antigen as the principal morphogenetic molecule. Nevertheless, it is clear that if H-Y antigen(s) have a role in testicular differentiation, the interaction of many other factors must also contribute to the final outcome.

Addendum

After completion of this chapter it has come to the attention of the author that: (1) the claim concerning trans-sexualism and H-Y_S antigen expression in humans has been withdrawn (Cleve, 1981); and (2) it has been reported that H-Y_C may be detected by H-Y-specific cytotoxic T lymphocytes on the fibroblasts, but not on concanavalin A-stimulated lymphocytes, of XO female mice (Koo *et al.*, 1983).

REFERENCES

Adinolfi, M., Polani, P. and Zenthon, J. (1982), *Human Genet.*, **61**, 1–2.

Akram, H. and Weniger, J.-P. (1968), *Arch. Anat. Microsc. Morphol. Exp.*, **57**, 369–378.

Andrews, P.W. and Wachtel, S.S. (1979), *Transplantation*, **27**, 43–44.

Bacon, L.D. (1970), *Transplantation*, **10**, 124–126.

Bacon, L.D. and Craig, J.V. (1966), *Poultry Sci.*, **45**, 1066–1067.

Bacon, L.D. and Craig, J.V. (1969), *Transplantation*, **7**, 387–393.

Bailey, D.W. (1971), *Transplantation*, **11**, 426–429.

Bailey, D.W. and Hoste, J. (1971), *Transplantation*, **11**, 404–407.

Benhaim, A., Gangnerau, M.-N., Bettane-Casanova, M., Fellous, M. and Picon, R. (1982), *Differentiation*, **22**, 53–58.

Bennett, D. and Boyse, E.A. (1973), *Nature (London)*, **246**, 308–309.

Bennett, D., Boyse, E.A. and Old, L.J. (1972), in *Cell Interactions; Third Lepetit Colloquium* (L.G. Silvestri, ed.), North-Holland, Amsterdam, pp. 247–263.

Bennett, D., Boyse, E.A., Lyon, M.F., Mathieson, B.J., Scheid, M. and Yanagisawa, K. (1975), *Nature (London),* **257,** 236–238.
Bennett, D., Mathieson, B.J., Scheid, M., Yanagisawa, K., Boyse, E.A., Wachtel, S.S. and Cattanach, B.M. (1977), *Nature (London),* **265,** 255–257.
Bernstein, R., Koo, G.C. and Wachtel, S.S. (1980), *Science,* **207,** 768–769.
Bernstein, S.E., Silvers, A.A. and Silvers, W.K. (1958), *J. Natl. Cancer Inst.,* **20,** 577–580.
Beutler, B., Nagai, Y., Ohno, S., Klein, G. and Shapiro, I. (1978), *Cell,* **13,** 509–513.
Bevan, M.J. (1975), *J. Exp. Med.,* **142,** 1349–1364.
Billingham, R.E. and Hings, I.M. (1981), *Human Genet.,* **58,** 9–17.
Billingham, R.E. and Silvers, W.K. (1959), *Transplant. Bull.* **6,** 399–403.
Billingham, R.E. and Silvers, W.K. (1960), *J. Immunol.,* **85,** 14–26.
Billingham, R.E., Hodge, B.A. and Silvers, W.K. (1962), *Proc. Natl. Acad. Sci. U.S.A.,* **48,** 138–147.
Billingham, R.E., Silvers, W.K. and Wilson, D.B. (1965), *Proc. R. Soc. London, Ser. B,* **163,** 61–89.
Bittner, J.J. (1932), *Am. J. Cancer,* **16,** 322–332.
Bloom, S.E. (1974), *BioScience,* **24,** 340–344.
Bricarelli, F.D., Fraccaro, M., Lindsten, J., Müller, U., Baggio, P., Carbone, L.D.L., Hjerpe, A., Lindgren, F., Mayerová, A., Ringertz, H., Ritzén, E.M., Rovetta, D.C., Sicchero, C. and Wolf, U. (1981), *Human Genet.,* **57,** 15–22.
Bunker, M.C. (1966), *Can. J. Genet. Cytol.,* **8,** 312–327.
Cattanach, B.M., Pollad, C.E. and Hawkes, S.G. (1971), *Cytogenetics,* **10,** 318–337.
Celada, F. and Welshons, W.J. (1963), *Genetics,* **48,** 139–151.
Chai, C.K. (1968), *Transplantation,* **6,** 689–693.
Chen, H.D. and Silvers, W.K. (1982), *J. Immunol.,* **128,** 2044–2048.
Cleve, H. (1981), in *Protides of the Biological Fluids* (Peeters, H., ed.), Vol. 29, Pergamon Press, Oxford, pp. 3–12.
Davis, R.M. (1981), *J. Med. Genet.,* **18,** 161–195.
de la Chapelle, A., Koo, G.C. and Wachtel, S.S. (1978), *Cell,* **15,** 837–842.
Eicher, E.M. (1982), *Prospects for Sexing Mammalian Sperm,* (R.P. Amann and G.E. Seidel, eds), Colorado Associated University Press, Boulder, Colorado, pp. 121–135.
Eicher, E.M., Washburn, L.L., Whitney, J.B. and Morrow, K.E. (1982), *Science,* **217,** 535–537.
Eichwald, E.J. and Davidson, N. (1968), *Folia Biol. (Prague),* **6,** 89–93.
Eichwald, E.J. and Lustgraaf, E.C. (1961), *J. Natl. Cancer Inst.,* **26,** 1395–1403.
Eichwald, E.J. and Silmser, C.R. (1955), *Transplant. Bull.,* **2,** 148–149.
Eichwald, E.J. and Wetzel, B. (1965), *Transplantation,* **3,** 583–585.
Eichwald, E.J., Silmser, C.R. and Wheeler, N. (1957), *Ann. N.Y. Acad. Sci.,* **64,** 737–740.
Eichwald, E.J., Silmser, C.R. and Weissman, I. (1958), *J. Natl. Cancer Inst.,* **20,** 563–575.
Engel, W., Pfäfflin, F. and Wiedeking, C. (1980), *Human Genet.,* **55,** 315–319.
Engel, W., Klemme, B. and Schmid, M. (1981a), *Differentiation,* **20,** 152–156.
Engel, W., Klemme, B. and Ebrecht, A. (1981b), *Human Genet.,* **57,** 68–70.
Engelstein, J.M. (1967), *Proc. Soc. Exp. Biol. Med.,* **126,** 907–912.

Epstein, C.J., Smith, S. and Travis, B. (1980), *Tissue Antigens,* **15,** 63–67.
Erickson, R.P. (1977), *Nature (London),* **265,** 59–61.
Evans, E.P., Burtenshaw, M.D. and Cattanach, B.M. (1982), *Nature (London),* **300,** 443–445.
Feldman, M. (1958), *Transplant. Bull.,* **5,** 15–16.
Fellous, M., Günther, E., Kemler, R., Wiels, J., Berger, R., Guénet, J.L., Jakob, H. and Jacob, F. (1978), *J. Exp. Med.,* **148,** 58–70.
Fierz, W., Brenan, M., Müllbacher, A. and Simpson, E. (1982), *Immunogenetics,* **15,** 261–270.
Fisher-Lindahl, K. and Lemke, H. (1979), *Eur. J. Immunol.,* **9,** 526–536.
Gasser, D.L. and Shreffler, D.C. (1974), *Immunogenetics,* **1,** 133–140.
Gasser, D.L. and Silvers, W.K. (1971a), *J. Immunol.,* **106,** 875–876.
Gasser, D.L. and Silvers, W.K. (1971b), *Transplantation,* **12,** 412–414.
Gasser, D.L. and Silvers, W.K. (1972), *Adv. Immunol.,* **15,** 215–247.
Gasser, D.L., Silvers, W.K. and Wachtel, S.S. (1974), *Science,* **185,** 963.
Gilmour, D.G. (1967), *Transplantation,* **5,** 699–706.
Gittes, R.F. and Russell, P.S. (1961), *J. Natl. Cancer Inst.,* **26,** 283–303.
Goldberg, E.H., Boyse, E.A., Bennett, D., Scheid, M. and Carswell, E.A. (1971), *Nature (London),* **232,** 478–480.
Goldberg, E., Boyse, E.A., Scheid, M. and Bennett, D. (1972), *Nature (London) New Biol.,* **238,** 55–57.
Goldberg, E.H., Shen, F. and Tokuda, S. (1973), *Transplantation,* **15,** 334–336.
Goodfellow, P.N. and Andrews, P.W. (1982), *Nature (London),* **295,** 11–13.
Gordon, J.W. and Ruddle, F.H. (1981), *Science,* **211,** 1265–1271.
Gordon, R.D., Simpson, E. and Samelson, L.E. (1975), *J. Exp. Med.,* **142,** 1108–1120.
Gordon, R.D., Mathieson, B.J., Samelson, L.E., Boyse, E.A. and Simpson, E. (1976), *J. Exp. Med.,* **144,** 810–820.
Gore-Langton, R.E., Tung, P.S. and Fritz, I.B. (1983), *Cell,* **32,** 289–301.
Goulmy, E., Bradley, B.A., Lansbergen, Q. and van Rood, J.J. (1978), *Transplantation,* **25,** 315–319.
Greene, M.I., Benacerraf, B. and Dorf, M.E. (1980), *Immunogenetics,* **11,** 267–273.
Hall, J.L. and Wachtel, S.S. (1980), *Mol. Cell Biochem.,* **33,** 49–66.
Hall, J.L., Bushkin, Y. and Wachtel, S.S. (1981), *Human Genet.,* **58,** 34–36.
Haseltine, F.P., Genel, M., Crawford, J.D. and Breg, W.R. (1981), *Human Genet.,* **57,** 265–268.
Haseltine, F.P., Lynch, V.A., Van Dyke, D.L., Breg, W.R. and Franke, U. (1982), *Am. J. Med. Genet.,* **13,** 115–123.
Hauschka, T.S. (1955), *Transplant. Bull.,* **2,** 154–155.
Hauschka, T.S. and Holdridge, B.A. (1962), *Ann. N.Y. Acad. Sci.,* **101,** 12–22.
Hildemann, W.H. (1974), *Science,* **185,** 962–963.
Hildemann, W.H. and Cooper, E.L. (1967), *Transplantation,* **5,** 707–720.
Hildemann, W.H., Morgan, M. and Frautnick, L. (1970), *Transplant. Proc.,* **2,** 24–31.
Hirsch, B.B. (1957), *Transplant. Bull.,* **4,** 58.
Hoshino, K. and Moore, J.E. (1968), *Int. J. Cancer,* **3,** 374.
Hurme, M., Hetherington, C.M., Chandler, P.R., Gordon, R.D. and Simpson, E. (1977), *Immunogenetics,* **5,** 453–459.

Hurme, M., Chandler, P.R., Hetherington, C.M. and Simpson, E. (1978a), *J. Exp. Med.*, **147**, 768–775.
Hurme, M., Hetherington, CM., Chandler, P.R. and Simpson, E. (1978b), *J. Exp. Med.*, **147**, 758–767.
Johnson, L.L. (1982), *Immunogenetics*, **16**, 577–582.
Judd, K.P. and Trentin, J.J. (1971), *Transplantation*, **11**, 298–302.
Klein, E. and Linder, O. (1961), *Transplant. Bull.*, **27**, 457–459.
Koo, G.C. (1981), *Human Genet.*, **58**, 18–20.
Koo, G.C. and Varano, A. (1981), *Immunogenetics*, **14**, 183–188.
Koo, G.C., Wachtel, S.S., Krupen-Brown, K., Mittl, R.L., Breg, W.R., Genel, M., Rosenthal, I.M., Borgaonkar, D.S., Miller, D.A., Tantravahi, R., Schreck, R.R., Erlanger, B.F. and Miller, O.J. (1977a), *Science*, **198**, 940–942.
Koo, G.C., Wachtel, S.S., Saenger, P., New, M.I., Dosik, H., Amarose, A.P., Dorus, E. and Ventruto, V. (1977b), *Science*, **196**, 655–666.
Koo, G.C., Goldberg, C.L. and Shen, F.W. (1979), *J. Exp. Med.*, **150**, 1028–1032.
Koo, G.C., Tada, N., Chaganti, R. and Hammerling, U. (1981), *Human Genet.*, **57**, 64–67.
Koo, G.C., Reidy, J.A. and Nagamine, C.M. (1983), *Immunogenetics* **18**, 37–44.
Kozelka, A.W. (1932), *J. Exp. Zool.*, **61**, 431–482.
Krco, C.J. and Goldberg, E.H. (1976), *Science*, **193**, 1134–1135.
Králová, J. and Démant, P. (1976), *Immunogenetics*, **3**, 583–594.
Krohn, P.L. (1958), *Transplant. Bull.*, **5**, 126–128.
Liew, F.Y. and Simpson, E. (1980), *Immunogenetics*, **11**, 255–266.
Little, C.C. (1941), in *The Biology of the Laboratory Mouse* (G.D. Snell, ed.), Blakiston, Philadelphia, pp. 279–309.
Lyon, M.F. and Hawkes, S.G. (1970), *Nature (London)*, **227**, 1217–1219.
McCarrey, J.R., Abplanalp, H. and Abbott, U.K. (1981), *J. Hered.*, **72**, 169–171.
McLaren, A. (1976), *Mammalian Chimaeras*, Cambridge University Press, London.
McLaren, A. and Monk, M. (1982), *Nature (London)*, **300**, 446–448.
Melvold, R.W., Kohn, H.I., Yerganian, G. and Fawcett, D.W. (1977), *Immunogenetics*, **5**, 33–41.
Michie, D. and McLaren, A. (1958), *Transplant. Bull.*, **5**, 17–18.
Miller, L. (1962), *Transplant. Bull.*, **30**, 147–149.
Miller, R.A. (1938), *Anat. Rec.*, **70**, 155–189.
Mullen, Y. and Hildemann, W.H. (1972), *Transplantation*, **13**, 521–529.
Müller, U. and Urban, E. (1981), *Cytogenet. Cell Genet.*, **31**, 104–107.
Müller, U. and Wolf, U. (1979), *Differentiation*, **14**, 185–187.
Müller, U., Aschmoneit, I., Zenzes, M.T. and Wolf, U. (1978a), *Human Genet.*, **43**, 151–157.
Müller, U., Siebers, J.W., Zenzes, M.T. and Wolf, U. (1978b), *Human Genet.*, **45**, 209–213.
Müller, U., Zenzes, M.T., Bauknecht, T., Wolf, U., Siebers, J.W. and Engel, W. (1978c), *Human Genet.*, **45**, 203–207.
Müller, U., Wolf, U., Siebers, J.W. and Günther, E. (1979a), *Cell*, **17**, 331–335.
Müller, U., Zenzes, M.T., Wolf, U., Engel, W. and Weniger, J.-P. (1979b), *Nature (London)*, **280**, 142–144.

Müller, U., Guichard, A., Reyss-Brion, M. and Scheib, D. (1980), *Differentiation,* **16,** 129–133.

Nagai, Y., Ciccarese, S. and Ohno, S. (1979), *Differentiation,* **13,** 155–164.

Naji, A., Frangipane, L., Barker, C.F. and Silvers, W.K. (1981), *Transplantation,* **31,** 145–147.

Ohno, S. (1971), *Nature (London),* **234,** 134–137.

Ohno, S. (1976), *Cell,* **7,** 315–321.

Ohno, S., Christian, L.C., Wachtel, S.S. and Koo, G.C. (1976), *Nature (London),* **261,** 597–599.

Ohno, S., Nagai, Y. and Ciccarese, S. (1978), *Cytogenet. Cell Genet.,* **20,** 351–364.

Ohno, S., Nagai, Y., Ciccarese, S. and Iwata, H. (1979), *Recent Prog. Horm. Res.,* **35,** 449–476.

Oliver, R.T.D. (1974), *Eur. J. Immunol.,* **4,** 519–520.

Pechan, P., Wachtel, S.S. and Reinboth, R. (1979), *Differentiation,* **14,** 189–192.

Poláčková, M. (1969), *Folia Biol. (Prague),* **15,** 181–187.

Poláčková, M. (1970), *Folia Biol. (Prague),* **16,** 12–19.

Poláčková, A. and Vojtíšková, M. (1968), *Folia Biol. (Prague),* **6,** 93–100.

Puck, S.M., Haseltine, F.P. and Franke, U. (1981), *Human Genet.,* **57,** 23–27.

Rary, J.M., Cummings, D.K., Jones, H.W. and Rock, J.A. (1979), *J. Hered.,* **70,** 78–80.

Rosenfeld, R.G., Luzzatti, L., Hintz, R.L., Miller, O.J., Koo, G.C. and Wachtel, S.S. (1979), *Am. J. Hum. Genet.,* **31,** 458–468.

Sachs, L. and Heller, E. (1958), *J. Natl. Cancer Inst.,* **20,** 555–561.

Scheid, M., Boyse, E.A., Carswell, E.A. and Old, L.J. (1972), *J. Exp. Med.,* **135,** 938–955.

Shalev, A. and Heubner, E. (1980), *Differentiation,* **16,** 81–83.

Shalev, A., Berczi, I. and Hamerton, J.L. (1978), *J. Immunogenet.,* **5,** 303–312.

Shalev, A., Goldenberg, P.Z. and Heubner, E. (1980a), *Differentiation,* **16,** 77–80.

Shalev, A., Short, R.V. and Hamerton, J.L. (1980b), *Cytogenet. Cell Genet.,* **28,** 195–202.

Shapiro, M. and Erickson, R.P. (1981), *Nature (London),* **290,** 503–505.

Short, R.V. (1979), *Br. Med. Bull.,* **35,** 121–127.

Siebers, J.W., Peters, F., Zenzes, M.T., Schmidtke, J. and Engel, W. (1977), *J. Endocrinol.,* **73,** 491–496.

Silvers, W.K. and Wachtel, S.S. (1977), *Science,* **195,** 956–960.

Silvers, W.K. and Yang, S.L. (1973), *Science,* **181,** 570–572.

Silvers, W.K., Billingham, R.E. and Sanford, B.H. (1968), *Nature (London),* **220,** 401–403.

Silvers, W.K., Gasser, D.L. and Eicher, E.M. (1982), *Cell,* **28,** 439–440.

Simpson, E. (1982a), *Nature (London),* **300,** 404–406.

Simpson, E. (1982b), *Immunol. Today,* **3,** 97–106.

Simpson, E. and Gordon, R.D. (1977), *Immunol. Rev.,* **35,** 59–75.

Simpson, E., Brunner, C., Hetherington, C., Chandler, P., Brenan, M., Dagg, M. and Bailey, D.W. (1979), *Immunogenetics,* **8,** 213–219.

Simpson, E., Edwards, P., Wachtel, S., McLaren, A. and Chandler, P. (1981), *Immunogenetics,* **13,** 355–358.

Simpson, E., McLaren, A. and Chandler, P. (1982), *Immunogenetics,* **15,** 609–614.

Simpson, E., Chandler, P., Washburn, L., Bunker, H. and Eicher, E.M. (1983), *Differentiation* (in press).
Singh, L. and Jones, K.W. (1982), *Cell*, **28**, 205–216.
Snell, G.D. (1956), *Transplant. Bull.*, **3**, 29–31.
Špoljar, M., Eicher, W., Eiermann, W. and Cleve, H. (1981), *Human Genet.*, **57**, 52–57.
Stimpfling, J.H. and Reichert, A.E. (1971), *Transplantation*, **12**, 527–531.
Vojtíšková, M. and Poláčková, M. (1966), *Folia Biol. (Prague)*, **12**, 137–140.
Wachtel, S.S. (1977), *Immunol. Rev.*, **33**, 33–58.
Wachtel, S.S. (1983), *H-Y Antigen and the Biology of Sex Determination*, Grune and Stratton, New York.
Wachtel, S.S. and Ohno, S. (1979), *Prog. Med. Genet.*, **3**, 109–142.
Wachtel, S.S., Gasser, D.L. and Silvers, W.K. (1973a), *Science*, **181**, 862–863.
Wachtel, S.S., Gasser, D.L. and Silvers, W.K. (1973b), *Transplant. Proc.*, **5**, 295–298.
Wachtel, S.S., Koo, G.C., Zuckerman, E.E., Hammerling, U., Scheid, M.P. and Boyse, E.A. (1974), *Proc. Natl. Acad. Sci. U.S.A.*, **71**, 1215–1218.
Wachtel, S.S., Ohno, S., Koo, G.C. and Boyse, E.A. (1975a), *Nature (London)*, **257**, 235–236.
Wachtel, S.S., Koo, G.C. and Boyse, E.A. (1975b), *Nature (London)*, **254**, 270–272.
Wachtel, S.S., Koo, G.C., Breg, W.R., Elias, S., Boyse, E.A. and Miller, O.J. (1975c), *N. Engl. J. Med.*, **293**, 1070–1072.
Wachtel, S.S., Koo, G.C., Breg, W.R., Thaler, T.H., Dillard, G.M., Rosenthal, I.M., Dosik, H., Gerald, P.S., Saenger, P., New, M., Lieber, E. and Miller, O.J. (1976a), *N. Engl. J. Med.*, **295**, 750–754.
Wachtel, S.S., Koo, G.C., Ohno, S., Gropp, A., Dev, V.G., Tantravahi, R., Miller, D.A. and Miller, O.J. (1976b), *Nature (London)*, **264**, 638–639.
Wachtel, S.S., Basrur, P. and Koo, G.C. (1978), *Cell*, **15**, 279–281.
Wachtel, S.S., Koo, G.C., Breg, W.R. and Genel, M. (1980a), *Human Genet.*, **56**, 183–187.
Wachtel, S.S., Hall, J.L., Müller, U. and Chaganti, R.S.K. (1980b), *Cell*, **21**, 917–926.
Wachtel, S.S., Bresler, P.A. and Koide, S.S. (1980c), *Cell*, **20**, 859–864.
Weissman, I.L. (1973), *Transplantation*, **16**, 122–125.
Weniger, J.-P. (1963), *Arch. Anat. Microsc. Morphol. Exp.* **52**, 261–275.
Weniger, J.-P., Zeis, A. and Engel, W. (1981), *C.R. Hebd. Séances Acad. Sci. Ser. D.*, **292**, 303–305.
Wiberg, U. (1982), *Differentiation*, **21**, 206–208.
Wiberg, U., Mayerová, A., Müller, U., Fredga, K. and Wolf, U. (1982), *Human Genet.*, **60**, 163–166.
Wolf, U., Fraccaro, M., Mayerová, A., Hecht, T., Maraschio, P. and Hameister, H. (1980a), *Human Genet.*, **54**, 149–154.
Wolf, U., Fraccaro, M., Mayerová, A., Hecht, T., Zuffardi, O. and Hameister, H. (1980b), *Human Genet.*, **54**, 315–318.
Zaalberg, O.B. (1959), *Transplant. Bull.*, **6**, 433–435.
Zaborski, P., Dorizzi, M. and Pieau, C. (1982), *Differentiation*, **22**, 73–78.
Zeiss, I.M., Nisbek, N.W. and Heslop, B.F. (1962), *Transplant. Bull.*, **30**, 49–51.
Zenzes, M.T., Müller, U., Aschmoneit, I. and Wolf, U. (1978a), *Human Genet.*, **45**, 297–303.

Zenzes, M.T., Wolf, U., Günther, E. and Engel, W. (1978b), *Cytogenet. Cell Genet.*, **20,** 365–372.
Zenzes, M.T., Wolf, U. and Engel, W. (1978c), *Human Genet.*, **44,** 333–338.
Zenzes, M.T., Urban, E. and Wolf, U. (1980), *Differentiation*, **17,** 121–126.
Zinkernagel, R.M. and Doherty, P.C. (1974), *Nature (London)*, **251,** 547–548.

Index